American Masculinity
under Clinton

Toby Miller
General Editor

Vol. 7

PETER LANG
New York • Washington, D.C./Baltimore • Bern
Frankfurt am Main • Berlin • Brussels • Vienna • Oxford

BRENTON J. MALIN

American Masculinity under Clinton

Popular Media and the Nineties "Crisis of Masculinity"

PETER LANG
New York • Washington, D.C./Baltimore • Bern
Frankfurt am Main • Berlin • Brussels • Vienna • Oxford

Library of Congress Cataloging-in-Publication Data

Malin, Brenton J.
American masculinity under Clinton: popular media
and the nineties "crisis of masculinity" / Brenton J. Malin.
p. cm. — (Popular culture and everyday life; vol. 7)
Includes bibliographical references and index.
1. Masculinity—Political aspects—United States.
2. Masculinity in popular culture—United States.
3. Popular culture—Political aspects—United States.
4. Mass media and culture—United States. 5. Clinton, Bill.
6. United States—Politics and government—1993–2001. I. Title.
II. Series: Popular culture & everyday life; vol. 7.
HQ1090.3.M24   305.31'0973—dc22   2004022835
ISBN 0-8204-6806-1
ISSN 1529-2428

Bibliographic information published by **Die Deutsche Bibliothek.**
Die Deutsche Bibliothek lists this publication in the "Deutsche
Nationalbibliografie"; detailed bibliographic data is available
on the Internet at http://dnb.ddb.de/.

Cover design by Joni Holst

The paper in this book meets the guidelines for permanence and durability
of the Committee on Production Guidelines for Book Longevity
of the Council of Library Resources.

© 2005 Peter Lang Publishing, Inc., New York
275 Seventh Avenue, 28th Floor, New York, NY 10001
www.peterlangusa.com

Printed in the United States of America

Table of Contents

Acknowledgments

I could not have completed this project without the generous feedback, support, care, friendship, and love of the many people who have helped me throughout my past several years of writing and research. My mentors, colleagues, and students at four different universities and colleges—the University of Iowa, in Iowa City, Iowa; St. Olaf College, in Northfield, Minnesota; Allegheny College, in Meadville, Pennsylvania; and San Francisco State University—have been particularly inspiring and engaging, always encouraging me to explore ideas in interesting and provocative ways. I owe them each a great deal of thanks.

As this book originally grew out of my dissertation research at the University of Iowa, I owe a special thank you to my dissertation committee members: Ralph Cintron, Joy Hayes, John Durham Peters (who, in fact suggested that I make Bill Clinton a central focus for my discussion), and Eric Rothenbuhler; and I owe more thanks than I can express to my dissertation and graduate school advisor, Bruce Gronbeck, who read through and edited multiple versions of this manuscript, helping me to hone and focus my arguments throughout. Without the generous support of these five people, I could never have completed this project.

A number of others have read or responded to sections of this book; others have listened to and talked with me regarding this project—some briefly and some quite extensively—offering helpful feedback and support. I thank Christopher Bakken, Jay Baglia, Ann Bomberger, Marilyn Bordwell Delaure, Derek Buescher, Steven Burch, Karlyn Kohrs Campbell, Preston Coleman, Gordon Coonfield, Mark Cosdon, Daniel Crozier, Timothy Dayton, Susan Douglas, Daniel Emery, Larissa Faulkner, Anthony Fleury, Tere Garza, Dina

Gavrilos, Larry Grossberg, Robert Hellyer, Chul Heo, Wilfredo Hernandez, Mike Keeley, Robert MacDougall, Marty Marchitto, Michael Mazenac, David Martins, Dan Meinhardt, Lloyd Michaels, David Miller, Thomas Nakayama, Mark Neumann, Jennifer O'Donnell, Kyra Pearson, Valerie Peterson, Ken Pinnow, Barbara Reiss, Gil Rodman, Adam Roth, Erin Sahlstein, Carol Schrage, Steve Schwarze, Paige Schilt, Ishita Sinha Roy, John Sloop, Laura Smith, William Sonnega, Shi-Che Tang, Karla Tonella, Beth Watkins, and Ned Winsborough. In addition, Toby Miller, Damon Zucca, and others at Peter Lang have been especially helpful and patient as I have worked to finish this project. My parents, Ralph and Carole Malin, and my brother, Bryan, have supported me all along; I thank them for always encouraging my fabulations.

The members of the University of Iowa's Project on the Rhetoric of Inquiry engaged in several lively discussions of the case studies laid out in the following chapters—I thank Fred Antzak, David Depew, Les Margolin, and John Nelson in particular. Likewise, participants at the Situating the Comedy conference at Bowling Green State University; the 4to Congresso Internacional de las Americas, in Puebla, Mexico; the National Communication Association's Masculinity at the End of the Millennium preconference; and the Third International Crossroads in Cultural Studies Conference, at the University of Birmingham, UK, all offered helpful feedback on various portions of my writing. Portions of chapter 4 appeared previously as "Memorializing White Masculinity: The Late 1990s 'Crisis of Masculinity' and the 'Subversive Performance' of *Man on the Moon*" in the *Journal of Communication Inquiry* 27(3), 239–255. I thank its editors for permission to reprint it here.

Finally, I thank my students at the University of Iowa, St. Olaf College, Allegheny College, and San Francisco State University, who sustained my energy in working to complete this project. Their willingness to engage their contemporary culture in interesting and exciting ways has continuously benefited both my thinking and my spirit. Our conversations on contemporary media culture and, in particular, American masculinity have motivated and inspired my writing throughout. It is to them that this book is dedicated.

Introduction

In the midst of the campaign for a 2004 Democratic presidential candidate, Bill Maher asked of John Kerry, "Are we really ready for our first metrosexual President?" Making one of his typical *Politically Incorrect* jokes—this one about Kerry's alleged Botox treatment—Maher nonetheless hit upon several ideas that play a central role within the following chapters. First, like Maher's comments, this book is about the connection between masculinity and politics. As should be obvious from the title, this book explores masculinity during the Clinton era, looking both at general ideas about masculinity depicted primarily within popular film and television during this time, as well as at representations of Bill Clinton's masculinity itself. In doing so, I hope to unpack one part of the relationship between popular culture and presidential politics, showing how the Clinton campaign capitalized on particularly popular ideas of masculinity (and the so-called new man of the '90s), as well as how Clinton's opponents sought to use these masculine ideals against him. In this, I follow closely on the work of Susan Jeffords (1994), whose book on Reagan-era masculinity is one inspiration for my own work.

But this book is about politics in another sense as well—the sense generally associated with phrases such as "cultural politics." Here, as in the work of cultural studies more generally, politics goes beyond mere presidential or governmental politics to include a wider variety of ways in which power operates in people's everyday lives. Given this second sense, this book explores not only *masculinity and politics* but also *the politics of masculinity*, exploring some of the ways in which American conceptions of masculinity work to organize cultural meanings and resources in ways that tend to empower some groups of people and to disempower others. In this context, my work here focuses on the interrelationship between cultural conceptions of masculinity and cultural conceptions of race, class, and sexuality, showing how these various identities interpenetrate one another and some of the consequences of these relationships.

All of these discussions are grounded in another idea implicit within Bill Maher's statement above: that, as a culture, we come to particular, shared understandings about appropriate and inappropriate expressions of masculinity (which John Kerry apparently violated by having a cosmetic procedure performed on his face). This idea is itself based on at least two other assumptions that are equally important to my following discussions. First, putting basic biology aside, the position to which I subscribe holds that the general assumptions about what makes someone masculine of feminine are largely culturally and historically determined. While twenty years ago it would most likely have seemed unthinkable for a police officer to get a makeover by a team of gay fashion consultants, at the beginning of the twenty-first century such an event can garner high ratings on *Queer Eye for the Straight Guy*. Similarly, while for most of the twentieth century it was considered unmanly for men to cry, at other points in time Western culture has celebrated manly tears as heroic (Lutz, 1999). Likewise, in his powered wig and pantaloons, George Washington would hardly have seemed a manly candidate for a mid-twentieth-century election. What is considered manly at one moment might be considered taboo at another, and vice versa. Such cultural changes demonstrate that masculinity is a historically and culturally contingent idea that manifests in different ways at different periods of time.

Second, I assume that we can read and decipher these basic assumptions about masculinity at different periods in time, and that these basic assumptions have real consequences for the people living during that cultural moment. Here, I'm suggesting that the cumulative meanings of manhood presented at a given time amount to a particular masculine zeitgeist or "structure of feeling" (Williams, 1977) that can be felt throughout the general culture. The conceptions of masculinity presented in films, television programs, advertising, popular music, and so forth both reflect and create this culture. Given that the creators of these programs try to capitalize on particular trends in order to win viewers and advertising revenue, the programs tend to reinforce the norms of the mainstream, dominant culture. In the process, they also create this culture by weaving it together in coherent packages for sale to viewers and advertisers alike.

These collections of meanings (along with other meanings at play within the culture, e.g., within politics, law, medicine, etc.) come together to form dominant visions of masculinity at particular moments of time. These are dominant in the sense that they become widely circulated and widely celebrated, not because they are the only visions of masculinity or the only ideals about masculinity at play at that time. There are certain to be competing visions, though those that seem too far out of step with the mainstream culture aren't as likely to make it onto the film or television screen, except as the subject of jokes and derision. Likewise, by talking about a dominant conception of masculinity, I don't intend to suggest that this is an idea that every man at a particular time lives up to, or even aspires to. I feel certain that there are some men today who do wear powdered wigs and pantaloons (and in fact I seem to remember Adam Ant and some other rock stars doing so in the '80s). Still, if they decide to run for president—or even get most mainstream jobs—they are likely to have a difficult time.

Drawing these last two points together—that masculinity is culturally and historically determined, and that given points in time have dominant conceptions of manhood with real consequences for the people of that time—brings me to a further elaboration: *masculinity is an invisible but very real social construction*. Masculinity is invisible in the sense that we often tend not to think about dominant identities—masculinity, whiteness, heterosexuality—as identities (an idea I take up more fully in chapter 1). But it's also invisible in the sense that every structure of feeling largely seems invisible or, in the least, taken for granted. On this, we might turn to one of Karl Marx's famous statements on ideology, which, as Marx formulates it, is a similar structure operating in people's lives. According to Marx: "*Sie wissen das nicht, aber sie tun es*" ("they do not know it, but they are doing it").[1] The practices of masculinity are quite visible in the sense that they are consistently represented in such things as films, television programs, and popular music, as well as in particular men's behaviors. However, these are seldom recognized as a coherent picture of masculinity; men (in both real life and the movies) are supposed to act masculine even if they don't know precisely what masculinity means.

To call masculinity a *very real social construction* is simply to reiterate the point I've already made above. Even though our ideas about manhood are in many ways a fiction—made up for a culture at a particular point in time—that does not mean that they are any less real for the people living through and experiencing them. Too many young boys have no doubt felt the wrath of this too real social construction on the playground and elsewhere. Howard Dean felt this as well during his 2004 bid to become the Democratic presidential candidate, as his quickly infamous "shouting" speech following the Iowa caucuses was seen as too exuberant and excited for a presidential candidate (apparently no one told him the rules; they simply expected him to follow them). While our ideas about masculinity are not the only determining factor, they are central to a number of important decisions we make on a daily basis: from what kinds of movies to watch to how to raise our children and from what sorts of jobs to take to whom to elect as president.

This book attempts to untangle some of these questions by looking closely at a number of different elements of the popular culture of the 1990s. By doing so, I work to capture and investigate one particular masculine zeitgeist. I've picked this particular decade not only because it's filled with interesting figures of masculinity—Bill Clinton in particular—but also because a number of scholarly and popular writers begin during this time to make masculinity a subject of conversation (positions that I lay out more fully in chapter 1). It is from these various writers that I borrow the phrase "crisis of masculinity"—a phrase whose origins I explore in chapter 1 and whose various meanings are central to this book overall. Likewise, terms such as "'90s guy," "new man," and "SNAG" ("Sensitive New Age Guy") abound in much of this popular and academic literature. I use these terms with some caution. I don't necessarily trust the decade—the '90s—as a measurement, and don't necessarily think there is anything particularly "new" about this new '90s man. Still, as I discuss in the next chapter, there are reasons to look at the '90s, most notably the fact that people feel compelled to posit such terms to describe this decade of American manhood in the first place. Much of my project aims at unpacking and interrogating these terms and using them to understand the ways in which the meanings of masculinity are negotiated at particular points in time and within the '90s in particular.

Because masculinity is a social construction, every discussion of manhood stakes a claim in its potential meaning. This book is no different. I have explored movies and television programs that help me interrogate and understand our cultural conceptions of masculinity; I have also left many interesting and important images out of my discussion. Again, my aim is not to delimit once and for all what masculinity did or should have meant for the '90s. Rather, I've attempted to offer one story of the '90s that sheds a light on a series of images and problems, not the least of which are the discussions surrounding President Clinton's campaigning, election, reelection, affair, and subsequent court hearings. My aim is to help open up masculinity as a subject of discussion and thus to help make it more visible. In this regard, this book is a work of both media and cultural criticism and should perhaps be viewed as much as an example of anthropology as it should a piece of literary or filmic analysis. I see the various popular cultural artifacts I explore as windows into the feelings, desires, and anxieties of the American culture of the 1990s. My various chapters seek to describe and interpret these artifacts, helping to explore their visions of masculinity through the lenses of gender, cultural, and media theory. In encouraging readers to see masculinity in a variety of popular cultural and political discourses, I hope to help people think more critically about the masculine ideals that permeate much of our culture and the potential consequences of these ideals.

The term *metrosexual*, used by Bill Maher above, illustrates a final idea of importance to my discussion here, that our conceptions of manhood have a very real economic importance within our larger culture. Masculinity, as an artifact of popular culture, is used to sell television programs, movies, automobiles, stereos, and nearly every other imaginable consumer good (not to mention political candidates, laws, wars, etc.). The idea of the metrosexual, which has developed in the early part of this century to describe men preoccupied with their appearance, has created a marketing windfall for companies attempting to sell male cosmetics, shaving accessories, and other products that have previously been seen as "unmanly." In the same way, Hollywood film companies, television networks, television producers, and advertisers all have vested interests in maintaining and circulating particular conceptions of masculinity.

It behooves us to pay careful attention to these images of manhood, particularly at a moment when the growth of media conglomerations means that fewer and fewer companies produce the various stories we tell ourselves as a culture. If our shared cultural conceptions of masculinity have a real consequence on our various thoughts and behaviors, it's important that we have a wide variety of images to think about, mull over, and negotiate as a public. The fewer companies producing our stories, the less variety of stories they are likely to tell, and the less various images of masculinity we are likely to see. As a result, it becomes all the more important to think critically about the images we do have. I hope that this book can help contribute to this dialogue and negotiation.

Note

1.   For further discussion of this quote, see Zizek (1989).

Chapter 1

Bill Clinton and the Crisis of Masculinity

> We are living at an important and fruitful moment now, for it is clear to men
> that the images of adult manhood given by the popular culture are worn
> out; a man can no longer depend on them.
>
> Robert Bly (1990, p. ix)

At a prayer breakfast on September 11, 1998, Bill Clinton made his
first public apology for his affair with Monica Lewinsky. "It is important to me that everybody who has been hurt know that the sorrow I
feel is genuine," he told the audience of clergy and reporters, "first
and most important, my family, also my friends, my staff, my Cabi-
net, Monica Lewinsky and her family, and the American people."
Continuing his apology, Clinton offered his "hope that with a broken
spirit and a still strong heart, I can be used for greater good for we
have many blessings and many challenges and so much work to do."
His self-proclaimed "broken spirit" and "still strong heart" served as
a testament to the conflicted ideas and identities that had followed
Clinton throughout his presidency. Broken yet strong, sensitive but
tough, Clinton was the model of a conflicted masculinity characteristic of the '90s. Sensitive to our pain, but tough on crime; wealthy
graduate of Yale, but down-home Arkansas boy; Clinton's persona
remained a bundle of conflicts that variously embraced and over-
turned different stereotypes of American masculinity. Clinton's masculinity was thoroughly conflicted—embracing a kind of new,
sensitive, nontraditional masculinity at the same time that it sought to
demonstrate a powerful, thoroughly established sense of "real

American manhood," the sort conventionally depicted in advertisements for pickup trucks by Ford, Dodge, and Chevy.

As a '90s man, Bill Clinton was not alone in this presentation of
conflicted, oddly negotiated American masculinities. While the '80s
was prolific with big muscled killers like Arnold Schwarzenegger,
who fired their guns indiscriminately to prove their own and America's massive strength, the typical '90s hero seemed to be more sensitive and more politically correct. Schwarzenegger calms down and
becomes a loving kindergarten teacher for *Kindergarten Cop*, though
his ability to get the bad guy (as he ultimately does, despite his loving, teacherly attitude) makes clear that he can still kick ass if he
needs to. While confused, strangely contradictory images of masculinity are a common feature in the history of American manhood,
these '90s men seem to have an importantly different tenor. Offering
various departures from traditional masculinity, a host of America's
'90s men seemed caught up in contemporary arguments critiquing
the heterosexist, patriarchal, classist, and racist values traditionally
underwriting the standard picture of the "real American man." Like
Bill Clinton, these popular '90s men depict a conflicted masculinity
that both embraces and puts aside a variety of masculine stereotypes.

Such conflicted examples provide evidence of the '90s "crisis of
masculinity," a set of challenges to traditional masculinity identified
by both popular and academic sources. With arguments regarding the
identity politics of race, class, gender, and sexuality working to critique the standards on which traditional masculinity had been built,
the notion of a true, real manhood underwent particular challenges.
While this critique of traditional manhood is at least as old as the
counterculture of the '60s, this sense of crisis seemed to gain a particular cultural currency in the '90s. Robert Bly's popular book *Iron
John*, for instance, was published in 1990, the same year Bill
McCartney's men's group, the Promise Keepers, held their first meeting. Similarly, such magazines as *The Nation*, *Los Angeles Times Magazine*, and *New York Times Book Review* variously took up this masculine
crisis, essays such as "The Prejudice Against Men" (Marin, 1991),
"The Trouble with Male Bashing" (Stillman, 1994), and "What do
men want? A reading list for the male identity crisis" (Shweder, 1993)
all pointing to this sense of a crisis in '90s masculinity.

A number of gender scholars throughout academia (Kimmel,
1996, 1993; Jeffords, 1994; Connell, 1993; Messner, 1993; Rotundo,
1993; Segal, 1993; Miles, 1989; Halberstam, 1998; Faludi, 1999) have

discussed this crisis of masculinity as well, excited by the ways in which this crisis might challenge traditional masculine values and open up discussions of alternative views of masculinity. R. W. Connell, for instance, asserts that "the fact that conferences about 'masculinities' are being held is significant, [for] twenty-five years ago no one would have thought of doing so." Connell (1993) cites the "advent of Women's Liberation at the end of the 1960's," "the growth of feminist research on gender and 'sex roles,'" and the "advent of Gay Liberation and the developing critique of heterosexuality of lesbians and gay men" (pp. 598–599), as factors in this contemporary critique of masculinity. Lynne Segal supports this idea, claiming that "anti-sexist men" have encouraged a "less traditional masculinity" by "sharing in some of the joys of traditional 'femininity.'"[1] Calling attention to the ways in which the crisis of masculinity has upset traditional maleness, these scholars see positive possibilities in crisis-of-masculinity discourses—discourses that potentially disrupt these traditional masculine ideals.

This celebration of the possibilities opened up by the crisis of masculinity makes sense when coupled with notions of masculinity's traditionally invisible, unmarked status. As Judith Butler (1990) illustrates through her readings of Simone de Beauvoir, Luce Irigaray, and Monique Wittig, masculinity is conventionally conflated with the universal citizen and as such remains "unmarked."[2] "Gender" is associated with femininity whereas masculinity is granted an abstracted, universal quality.[3] In other words, masculinity proves "universally generalizable" (Kimmel, 1996, p. 4), such that men are not particular identities; they are simply citizens. This unmarked character is empowering for masculinity (in particular, white, middle-class, heterosexual masculinity) and disempowering for the others— the marked—alongside whom this unmarked identity is exercised. As Richard Dyer suggests, "the claim to power is the claim to speak for the commonality of humanity" (1997, p. 2); unmarked, universal bodies can do that. Marked bodies cannot. Because the discourses accompanying the '90s crisis of masculinity make masculinity an explicit topic of critique, they work against this traditional invisibility, and thus against a source of masculinity's traditionally oppressive power. Hence, Connell and others suggest, such a sense of a crisis of masculinity might help to dislodge the power of traditional maleness, opening up new possible conceptions of masculinity.

In the following chapters I interrogate this crisis of masculinity from a media and cultural studies perspective, attempting to understand the new possibilities presumably afforded by '90s masculinity. Looking at a variety of television programs and popular films, I explore the ways in which a set of '90s men negotiates this climate of '90s masculinity. On the one hand, I maintain with the masculinity scholars above that the challenges of the '90s open up new possibilities for American manhood, leaving behind traditional maleness in a variety of ways. By challenging the heterosexism, classism, racism, and sexism that have underwritten American manhood, crisis-of-masculinity discourses have indeed challenged masculinity's invisibility and encouraged a "less traditional masculinity." In the wake of these '90s discourses, Bill Clinton could express his sensitive side and Arnold Schwarzenegger needed not simply blow up buildings and people. Such changes have led scholars to identify a '90s "new man," who, in Susan Jeffords words, "can transform himself from the hardened, muscle-bound, domineering man of the eighties into the considerate, loving, and self-sacrificing man of the nineties" (1994, p. 153). Here, America seemed to get a nontraditional male hero who departs from the hypermasculine traditions of Reagan's '80s.

However, the following analyses demonstrate that the "possibilities" opened up by the "gender trouble" of the crisis of masculinity include the possibility of salvaging the disturbing traditions from which the '90s man seemed to diverge. In each of the following case studies (chapters 2 through 4), a supposedly new, '90s masculinity is accompanied by a concomitant anxiety associated with traditional male values. For instance, "sensitive men" are made hypersexual, highlighting their heterosexuality in order to stave off fears of homosexuality. "Lower-class" male heroes are given degrees of middle-class cultural capital, holding lower-class masculinity at bay at the same time that they appear to embrace it. In other words, the '90s men discussed in the following chapters do not simply use the "possibilities" of the crisis of masculinity to explore new male forms. Rather, these possibilities become avenues through which they offer a newly negotiated traditional maleness, one that reiterates traditional anxieties about manhood. Masculinity scholarship, I argue, needs a thorough exploration of these anxieties, if only to open up this new maleness for discussion, and, as much as possible, to understand and disrupt the mechanisms of power refigured within this new masculinity. Exploring a variety of bizarre, oddly conflicted images of '90s manhood, I attempt to demonstrate their tragicomic character, exploring the anxieties and emotional conflicts they wreak on themselves and others.

The next few sections introduce these ideas more fully. The first section offers a recent history of the crisis of masculinity, illustrating the discussions that led up to the '90s crisis of masculinity. The following section demonstrates one way in which this contemporary crisis has impacted the American cultural scene, exploring the controversies surrounding President Clinton's sexuality not so much as a crisis of the presidency but as part of the contemporary crisis of masculinity. Throughout, I work to demonstrate the ways in which a set of anxieties is articulated and rearticulated alongside the "new men" of the '90s. In the process, I suggest that in addition to exploring ideas of ideology, masculinity scholars should attend more carefully to "emotionology," the complex structures of feeling that these changing notions of masculinity negotiate, redefine, and reiterate (Stearns and Stearns, 1985; Williams, 1977).

The Crisis of Masculinity

> In an important sense there is only one complete unblushing male in America: a young, married, white, urban, northern, heterosexual Protestant father of college education, fully employed, of good complexion, weight, and height, and a recent record in sports.... Any male who fails to qualify in any of these ways is likely to view himself—during moments at least—as unworthy, incomplete, and inferior; at times he is likely to pass and at times he is likely to find himself being apologetic or aggressive concerning known-about aspects of himself he sees as undesirable. (Goffman, 1963, p. 128)[4]

When Erving Goffman wrote this in 1963 he was identifying the sort of mainstream masculinity present in such television programs as *Leave It to Beaver*, in which Ward Cleaver evinced a stereotypically masculine fatherhood that dominated the American cultural scene. While this unblushing male held prominence at the time, he also began to experience a variety of critiques. Betty Friedan's book, *The Feminine Mystique*, for instance, likewise published in 1963, can be read as both an exploration of '60s femininity and an implicit critique of its contemporary masculinity. With chapters such as "The Problem That Has No Name," and "The Crisis in Woman's Identity," Friedan's book challenges the place of women within her contemporary culture and, by implication, the place of their male counterparts. By questioning relationships between men and women, Friedan poses a poignant problem for her contemporary men, one that masculinity scholars take up in a variety of ways.

Myron Brenton's 1966 book, *The American Male,* addresses these issues by offering a relatively pro-feminist rereading of late '60s manhood.[5] His first chapter, "The Male in Crisis," echoes the phrasing Friedan uses to discuss '60s women. Patriarchy, Brenton asserts, has not only impacted negatively upon women but upon men as well:

> This book is about the plight of the contemporary American male. It's about the increasingly difficult choices he is having to make, about the multiplicity of demands he is having to meet, and, most of all, about the invisible straightjacket that still keeps him bound to antiquated patriarchal notions of what he must do or be in order to prove himself a man. (p. 15)

From Brenton's perspective, these rigid patriarchal notions of manhood are at the root of the contemporary masculine crisis. "The degree to which a man subscribes to the traditional masculine stereotypes dictates in part the attitude he has regarding women and the world he must share with them," Brenton comments later. "The more trapped he is by a rigid view of masculinity, the more dismayed he finds himself" (pp. 69–70). This "beleaguered male," as Brenton identifies him, comes to the "unrealistic and unscientific conclusion that we are indeed living in a twentieth-century matriarchy" (p. 70), one in which "the mark of the female is everywhere" (p. 69). Brenton sees his contemporary men caught in a rigid "masculinity trap" in which late '60s discussions about American women induce anxiety by challenging men's rigid gender roles. Deeply invested in patriarchy, his contemporary men react to the women's movement with anxiety and fear.

In the years since the publication of Friedan's *Feminine Mystique* and Brenton's *American Male,* a number of other thinkers have taken up these masculinity issues as well, writing from both relatively pro-feminist (Doyle, 1983; Fasteau, 1975; Farrell, 1974)[6] and antifeminist (Doyle, 1976; Farrell, 1993[7]; Goldberg, 1974) perspectives. In his 1974 book, *The Liberated Man,* for instance, Warren Farrell calls men to liberate themselves from the tyranny of patriarchy, reiterating Brenton's take on masculinity. In contrast, Richard Doyle's *The Rape of the Male* (1976) locates the crisis of masculinity within the American legal system, claiming that feminism has contributed to the mistreatment of men in divorce proceedings. James Doyle's 1983 book, *The Male Experience*, provides an early attempt at a "men's studies" textbook, working to address contemporary manhood in relatively pro-feminist terms. Often arguing from distinctly different perspectives, these pieces have maintained a discourse of masculine crisis throughout the several decades leading up to the '90s.

Following these works, Robert Bly's *Iron John*, one of the most notorious texts of '90s manhood, neither initiated this masculine crisis nor offered the first contemporary problematization of American manhood. Rather, *Iron John* seems but the most recent addition to this crisis of masculinity list. And yet, more so than these other works, Bly's book struck a nerve within his contemporary society. Besides spending 62 weeks on the *New York Times* best-seller list, in 1991 *Iron John* was North America's best-selling hardback nonfiction book. Why such popularity? Bly's work came at a crucial time in the growing crisis of masculinity discourse. More vocal than ever, from growing movements in support of women's rights to increasing conversations about alternative lifestyles,[8] a proliferation of contemporary discourses seemed to destabilize traditional American masculinity. Ian Miles (1989) summarizes these ideas, tracing this crisis to

> the relative decline of traditional class-based foci of political action, and the rise of new movements oriented towards 'consumption politics' (lifestyle, ecology, welfarism, and so on) as well as 'production politics' (employment, working conditions, wages and so on). Sexuality is an important issue in several of these movements, especially in those championing gay liberation and homosexual rights—which are clearly difficult to accord with traditional masculinity—as well as in women's movements. But many other movements also begin to call some established masculine practices into question, by proposing, for example, new relations between economic activities and natural environments, new divisions between paid and unpaid work, new ways of resolving conflicts in the nuclear age, and so on. (p. 52)

In the '90s these accumulating voices seemed to have reached a critical mass, gaining popular as well as academic attention. "Indeed," Lauren Berlant (1997) has argued, "today many formerly iconic citizens who used to feel undefensive and unfettered feel truly exposed and vulnerable. They feel anxious about their value to themselves, their families, their publics, and their nation. They sense that they now have *identities*, when it used to be just other people who had

them" (p. 2). Such is the situation that invited Bly and his contemporaries to reflect upon and negotiate the contemporary crisis of masculinity.

My discussion takes up masculinity at the post-Bly convergence
of these multiple discourses. Specifically, I explore the ways in which
this unblushing male has been transformed by these multiple voices,
considering how discussions of gender, race, class, and sexuality have
helped to refashion traditional American manhood in the '90s. In the
process of this investigation, I myself undertake an uncomfortable but
necessary negotiation of ideas such as "crisis" and "new" and "old"
masculinities. On the one hand, my arguments depend upon a certain
"newness" of the masculinity of the '90s, as well as on a recognition
of the rhetorical and discursive reality of the "crisis of masculinity."
Indeed, a portion of my arguments, as I have already suggested, seeks
to demonstrate the ways in which the masculinity of the '90s responds to the proliferating discourses of masculine crisis that have
come to dominate the '90s. Thus, in selecting '90s men to discuss, I
have intentionally selected ones that seem to suggest this sense of crisis, illustrating particular kinds of conflicts that echo the discourses of
masculine crisis laid out above.

While this might seem to suggest my own investment in proving
the newness or uniqueness of '90s masculinity, this would misread
my overall argument. By selecting '90s men who resonate with the
'90s crisis of masculinity, I seek to demonstrate the dynamic possibilities of crisis-of-masculinity discourses, possibilities that may just as
likely reinscribe as overturn dominant notions of masculinity. The
"new men" I discuss are paradoxical, as is the "'90s crisis of masculinity" itself, creating ways to both depart from, and salvage, Goffman's
unblushing male in America. In a similar way, my use of the term
"the '90s crisis of masculinity" should not suggest that a sense of
masculine crisis is either completely indicative of, or unique to, the
1990s. Unambiguously traditional notions of masculinity still exist in
the '90s, just as other notions of masculinity have existed at earlier
moments in time. Still, the proliferation of discourses of masculine
crisis and discourses identifying a specifically "'90s crisis of masculinity" deserve attention and analysis.

My discussion works to understand the variously conflicted masculinities of the '90s. Although I focus primarily on popular media
representations, Bill Clinton, the exemplar of conflicted masculinity,
makes his presence known throughout. Investigating a series of mas-

culine representations that variously defy traditional masculine ideals, the central chapters explore a set of anxieties at work across these interesting moments of masculinity, exploring anxieties of sexuality (chapter 2), class (chapter 3), and race (chapter 4), as they are articulated via a series of '90s men. Because each of these chapters explores a different set of men and a different concomitant anxiety, the cultural conflicts described in each is also slightly different. So, for instance, the troubled sensitivity of the little men in chapter 2 works differently than the troubled whiteness of Andy Kaufman and Jean-Luc Picard in chapter 4. Even so, these collections of masculinities share a move to stitch back together an unblushing male in America, evincing a particular thread that runs against the new maleness these men presume to take up. Together, I suggest throughout, these masculinities and their accompanying anxieties have important implications for the late '90s debates about President Clinton, as well as for contemporary men and the "others" against whom these anxieties are directed.

Sometimes a Cigar Is Just a Penis

From TR's rugged masculinity to Jimmy Carter's soft-spoken, sensitive maleness, a president's manhood can serve as a barometer of the nation's visions of masculinity more generally. The late '90s scandals surrounding President Clinton illustrate the '90s crisis of masculinity in exemplary fashion, demonstrating the profound manner in which this crisis of masculinity confronted the American public during this time. With the Clinton-Lewinsky controversy, this so-called masculine crisis was no longer a mere academic abstraction but a collection of anxieties explicitly being struggled over within popular culture. Obsessed with Clinton's sexuality, the model he sets for America's children, and his abilities to perform under pressure, discussions of Clinton's character indicate a sense of masculine, not merely presidential, crisis.

Of course, President Clinton's conflicted manhood was public record long before Monica Lewinsky's name became cliché. A January 1992 article in *Time* highlighted the conflicted character of a pre-presidential Clinton:

> Clinton remains something of an enigma, the more so since he often seems a
> bundle of contradictions: a visionary leader and a poor manager; a pro-
> pounder of bold programs and a waffler who talks on both sides of hot is-
> sues. All of which raises the insistent question: Is Clinton for real—not only
> as front runner but as man, as Governor, as candidate? (Church, 1992, p. 16)

Framing these contradictions explicitly as questions of manhood and
asking whether Clinton is "a real man," this article clearly evokes the
sense of masculine crisis noted above. In still other popular discourses
of the time, Clinton seems precariously positioned between redneck,
power-hungry patriarch, and sensitive man of the '90s, further illustrating his seemingly enigmatic masculinity. According to a 1992 article in the *Washington Post*, "He's Elvis Presley with a calculator on his
belt, an outsized candidate with a drawl as big as his brain" (Von
Drehle, 1992, p. 1). Here we see not one but two patriarchs, "Good Ol'
Billy, the affable, self-described 'redneck' who calls a barbecue a
'feedin','" and "Policy Wonk, who stores mountains of information in
a single scan," the first the backwoods clod, and the second the
boardroom mastermind. Both draw on hypermasculine images, but
with distinctly different stereotypes and connotations.

Still other writers focus on the '92 Clinton's sensitivity, whether
celebrating his emotional openness or noting the emotional turmoil
he suffers as a result of the presidential campaign. Discussing a picture of Clinton with his arm around his friend Paul Berry, one writer
commends "Clinton's initiative at congeniality," as well as his "comfort with man-to-man touch":

> Without running the risk of being considered "touchy-feely," Clinton is
> known as a hugger of men and women. Simple handshakes aren't enough
> for this man whose theme song easily could be borrowed from the cotton
> industry's: "The touch, the feel, the fabric of our lives".... He and Hillary
> routinely hold hands—a universal sign of equality and involvement, and
> maybe a concern about slippery airplane stairs—and recently the television
> news cameras caught Bill giving Hillary a light stroke upward on her cheek,
> a touching gesture half the female population would sell their souls to have
> the men in their lives do just once, and not when they have ketchup on their
> faces. (Mathias, 1992, p. 5)

Focusing on more negative aspects of Clinton's "sensitivity," a September 1992 article in the *New York Times* entitled "The 1992 Campaign: Anxious in his lead, Clinton fights to run the race his way"
stresses the anxiety Clinton exhibits relative to presidents past: "Part

of the ritual of presidential campaigns is the ceremonial boarding and unboarding of the plane, with the candidate walking past a gauntlet of reporters shouting questions on the issues of the day. A relaxed candidate will stop and take a few. Gov. Bill Clinton strides right on by these days" (Toner, 1992, p. 1). Stressing the various pressures Clinton has experienced,[9] the article concludes, "In Falls Church, and at every event in recent days, he talked of the number of days remaining, just over 50 now. But he did so almost grimly, warily, like a man who sees a finish line, but one so distant it gives little solace."

Seen variously as a redneck, a conniving politician, a sensitive hugger of men and women, and an anxiety-ridden struggler in search of solace, the Clinton of 1992 illustrates a masculinity conflicted in a variety of ways. Contributing to the conflict, a 1992 Republican campaign ad calls attention to pre-Lewinsky scandal Gennifer Flowers, casting Clinton as hypersexual womanizer: "'What really happened between Bill Clinton and Gennifer Flowers?' the ad says. 'Did he lie about their affair? Did he try a cover-up? Call [the number] and get to know Bill Clinton the way Gennifer Flowers did'" (Kurtz, 1992, p. 10). This man from Hope, Arkansas, indeed seems split between competing definitions of masculinity—cast simultaneously as insensitive and hypersensitive, as powerful politician and impotent draft dodger, as equal opportunity hugger and hypersexual player. The 1992 Clinton persona embodies the crisis of masculinity identified in the popular and academic discussions mentioned above.

And yet this early Clinton is relatively stable in comparison to the Clinton of the late '90s. In the wake of the Monica Lewinsky controversies, the Starr Report, and Clinton's taped testimony, the public manifestations of Clinton's conflicted masculinity are all too evident. An editorial in *Newsweek* published shortly after the sexually explicit Starr Report and Clinton testimony were released to the public calls Clinton "shameless yet compelling" (Alter, 1998, p. 45), illustrating this odd split within his persona. Like the earlier Clinton, the Lewinsky-era Clinton seems positioned paradoxically between *hyper* and *in*sensitivity. Evidencing the public's perceptions of Clinton's *in*sensitivity, enough Americans apparently felt his first admission of an "inappropriate relationship" with Monica Lewinsky was insincere—that he did not seem appropriately remorseful for his misdeeds, or for lying to the American public—to move Clinton to a series of more "heartfelt" apologies. "I was not contrite enough," Clinton later commented during his White House prayer breakfast. "I have re-pented."

While Clinton's earliest apologies were seen by many (including, later, Clinton himself) as insincere, insensitive word games, other dis-cussions of Clinton's persona seem to stress his *hyper*sensitivity— recalling the perceptions of Clinton's anxiety during the 1992 presi-dential campaign. A March article in the *New York Times* suggests that Clinton is "under siege in terms of his own psyche" (Bennett, 1998, p. 1), while a May 1998 editorial in the same newspaper claims to fur-ther recognize the profound emotional toll the Lewinsky scandal has taken on Clinton:

> In his first solo news conference since December, President Clinton gave his usual fluid performance, but it was obvious that he has paid a high price for his silence on the allegations about Monica Lewinsky. Most of Mr. Clinton's feelings came across through innuendo and body language. His repeated ducking of questions seemed defensive, argumentative and yet uncharacter-istically passive. Asked whether his behavior should even matter to Ameri-cans, the President all but shrugged and said he was 'in some ways the last person who needs to be having a national conversation about this.' For a man addicted to conversation, it was a revealing answer. (*New York Times*, 1998, p. 26)

Likewise, in an essay for *Newsweek*, former presidential advisor George Stephanopoulos provides readers with another glimpse into Clinton's psyche:

> As Bill Clinton and I walked into the Manchester Holiday Inn, a reporter handed me the faxed copy—it was blurry, but the headline was bold: They Made Love All Over Her Apartment. In a penthouse upstairs, Clinton read the piece, and as he moved through the paragraphs, he would seize on in-correct details and even managed a laugh when he found specifics he could disprove. But it was a nervous laugh; he was agitated, unsettled. (Stepha-nopoulos, 1998, p. 37)

Framed as insensitive and nonemotional in his apology to the public and yet hypersensitive and hyperemotional in these other instances, Clinton's conflicted character reached profound levels of contradic-tion—simultaneously *hyper* and *hypo*masculine.

Nowhere is this conflicted character more profound, however, than within the post-Starr Clinton delivered to us via the Starr Report and the four-hour videotape of Clinton's testimony released by Starr to the public. On the one hand, Clinton's perceived insensitivity shines through as clearly here as in any of the discussions above. "In an e-mail to another friend in early 1997," the report explains, "Monica Lewinsky wrote: 'I just don't understand what went wrong, what happened? How could he do this to me? Why did he keep in contact with me for so long and now nothing, now when we could be together?'" Likewise, the various sexual escapades recounted in the report seem to frame Clinton as the embodiment of a stereotypical hypersexual masculinity, further demonstrating this hypermasculine emphasis. At the same time, the Starr Report gives us a picture of a profoundly hypersensitive, vulnerable Clinton, who "once confided in Ms. Lewinsky that he was uncertain whether he would remain married after he left the White House," and who wore Monica's ties to keep her close to his heart. The Starr Report's infamous cigar story serves as a vignette of Clinton's paradoxical masculinity. Both a symbol of Clinton's sexuality and potent hypermasculinity (indeed, the cigar is a staple symbol of manhood), it also serves to demonstrate his impotence—the prosthesis he requires to consummate this sexual act. An earlier newspaper article had nicknamed Clinton "the Viagra Kid" (Rich, 1998) apparently foreshadowing this conflict between potency and impotency within the Clinton persona.[10]

Cast as both hypermasculine and impotent, the cigar-wielding Clinton personifies this '90s crisis of masculinity, illustrating its profound public implications. Might the late '90s debates regarding Clinton's ability to govern reflect not merely upon public perceptions of the presidency but upon public notions of American manhood? In what ways might these anxieties resonate with anxieties of sexuality, race, class, or other tensions implicated within this crisis of masculinity? The following chapters explore this crisis, attempting to understand how these conflicted masculinities come to be articulated within particular deployments of masculinity, whether those are portrayals of the president, portrayals of television fathers, or portrayals of action heroes. In each case, a better understanding of these conflicted masculinities, and the anxieties implicit within them, helps to understand public crises of masculinity, like that in which Clinton becomes involved, as well as the ongoing construction of masculinity more generally.

From Robert Bly to Cultural Studies

While Robert Bly may find the "images of adult manhood given by popular culture" to be "worn-out," my analysis seeks to demonstrate the ways in which these images maintain active negotiations of contemporary notions of maleness and the contemporary crisis of masculinity. Indeed, Bly's claim that "a man can no longer depend on" these images only reinforces the assertions made throughout these various chapters. Bly's own fear that he can no longer depend upon these images demonstrates the very anxieties at work within this contemporary sense of crisis as well as within the contemporary '90s men analyzed in chapters 3 through 4. Bly's *Iron John*, like these other images, is a negotiation of these anxieties, attempting to both address changing conceptions of masculinity and to hold onto some masculinity past. In contrast my approach to masculinity seeks to "mark" maleness, to highlight its constructedness as well as the anxieties at work within it. This is important not only in combating masculinity's "invisibleness" but also in bringing to the fore the prejudices and power imbalances at work within this invisibility. The '90s men explored here have won their new sensitivity at the expense of those victim to these "new male" fears and anxieties. Exploring these anxieties both appreciates the psychical force that is masculinity and seeks to reconcile the oppressive powers emanating from these anxieties.

This study is hardly intended to be exhaustive, either in its theoretical understandings of masculinity or in its exploration of images and notions of maleness. Of the myriad of possible '90s men I have, indeed, selected but a small number for investigation. Rather than positing a definitive picture of '90s masculinity, however, as if such a picture were possible, I attempt to illustrate the ways in which anxieties surrounding maleness work to both transform and reproduce the unblushing male in America. Although these '90s men may be, at times, markedly different from their masculine forefathers, they nonetheless illustrate much of their same fears, though often in a more subtly new male form. Thus, this tale of the '90s man is one of contradiction, anxiety, and tragedy, both for himself, and for those at whose expense he wins his new sensitivity. Only by exploring these anxieties can we begin to understand the emotional power of gender construction and the psychical powers of masculinity more generally. On the one hand, this serves as a sort of parodic repetition of contemporary maleness (see Butler, 1990), attempting to mark masculinity by holding up its darkly comic, tragic, contradictory, negotiations of contemporary culture. At the same time, it seeks to explore the power imbalances implicit within, and reinforced by, these anxieties, illustrating the subtle oppressiveness of these new '90s men, not only for the men themselves but for those others against whom they flex their new manhood.

In these ways, the tale of American manhood told here is, admittedly, a biased one—but one narrative selected from among the many possible stories of '90s maleness. Indeed, to hold that a single, simple narrative of contemporary maleness can be told would be to reinscribe the universally generalizable character that is the history of masculinity. Thus, my discussion here, like any discussion of the crisis of masculinity, does not merely address some objective notion of manhood at work in the American public. Indeed, from the Promise Keepers to Robert Bly to the various academic discussions of masculinity noted above, these explorations of masculinity and its crisis do not merely address the world "out there" but actively constitute this world, and thus this masculine crisis, in various ways as well. In short, critics, both popular and academic, discuss this crisis of masculinity in order to claim a stake in redefining manhood—the Promise Keepers hoping to reclaim a religiously conservative, traditional view, and Robert Bly an ahistoric mythical one.

Like these other discussions, my own discussion works to constitute this crisis in particular terms, inflected by particular perspectives, and aimed towards particular goals. Specifically, my readings are influenced by a meshing of American and British Cultural Studies, evidenced both in my methods of analysis and in the various theories on which I draw. According to American scholar James Carey (1977), while behavioral science typically seeks to "explain phenomena by assimilating them to either a functional or causal law" (p. 417), a cultural studies approach

> does not seek to explain human behavior, but to understand it. It does not seek to reduce human action to underlying causes or structures, but to interpret its significance. It is not an attempt to predict human behavior, but to diagnose human meanings. (p. 418)

Influenced by this American perspective to which Carey subscribes, my exploration of masculinity seeks just this type of meaning. By constructing readings of a variety of deployments of masculinity, these chapters work "to interpret the interpretations" (p. 421) of '90s masculinity drawn from a series of '90s sources. At the same time, influenced by a British tradition of cultural studies, drawn especially from the work of the Birmingham Center for Cultural Studies, my readings illustrate a concern for race, class, and gender that is emblematic of this British tradition. While some scholars of American cultural studies work in a more formal, literary tradition, one that is not explicitly political (see Campbell, 1991, for instance), my own approach takes power and cultural politics as a central focus, attempting to diagnose and interpret cultural anxieties regarding American masculinity and the power struggles at work within them.

Offering its own interpretive and politicized reading of contemporary masculinity, my analysis seeks to work against the stories of '90s manhood told by the likes of Kenneth Starr and the Promise Keepers. The Starr Report and the Promise Keepers both claim a stake in the contemporary crisis of masculinity—formulating this crisis for particular political and/or religious goals. Likewise, my discussion stakes its own claim in this crisis talk. Whereas the Promise Keepers and Kenneth Starr, however, seek to recover a masculinity past—the family values father, the good President—my own discussion seeks to disturb masculinity as a whole, parodying and demystifying the process of masculine ritualization and its accompanying conceptions of masculinity. Likewise, whereas the Promise Keepers and Kenneth Starr attempt to leave implicit the anxieties on which this crisis rhetoric depends, my discussion works to open up these anxieties to debate and critique, making explicit the emotions and fears articulated through a variety of masculine representations. In so doing, my arguments and analyses work to understand the emotional meanings articulated within these images of masculinity, diagnosing a set of anxieties at work within the new man of the '90s—the bizarre, conflicted emotions that structure this new masculinity.

Notes

1.   Yet, Segal continues, "the increasing more 'feminine' traits in some men is mocked by continual production and popularity of all the old hard, assertive, violent images of masculinity—from 'Rocky' to 'Rambo,' from 'Top Gun' to 'Lethal Weapon'" (p. 634).

2. The nature of this "invisibility" is substantially different for Beauvoir, Irigaray, and Wittig, however, as Butler demonstrates in her first chapter. While Beauvoir "contends that the female body is marked within masculine discourse, whereby the masculine body, in its conflation with the universal, remains unmarked," Irigaray "suggests that both marker and marked are maintained within a masculinist mode of signification in which the female body is 'marked off,' as it were, from the domain of the signifiable" (p. 12). While Butler resists embracing any one of these positions too enthusiastically, later, in *Bodies That Matter* (1993), she does suggest a way of exploring this "unmarked" masculine body against, and alongside, a marked femininity upon which it depends. "Disavowed," she writes in the first chapter, "the remnant of the feminine survives as the *inscriptional space* of that phallogocentrism, the specular surface which receives the marks of a masculine signifying act only to give back a (false) reflection and guarantee of phallogocentric self-sufficiency, without making any contribution of its own" (p. 39). Likewise, she describes masculinity as "a figure of disembodiment, but one which is nevertheless a figure of a body, a bodying forth of a masculinized rationality, the figure of a male body which is not a body" (p. 49) indicating a sort of "unmarked," "universally generalizable" manliness. This disembodied, generalizable masculinity "requires that women and slaves, children and animals be the body, perform the bodily functions, that it will not perform" (p. 49).

3. See also Michael Kimmel (1996, 1993) and Michael Warner (1994). Warner, for instance, argues that when people participate in a public sphere, it is always as an abstracted, universal subject that is masculine and heterosexual.

4. See Kimmel (1996) for a discussion of Goffman's unblushing American male.

5. Brenton, however, addresses Friedan only in passing, commenting that "one of the lady neo-feminists, who has written extensively on the trapped-housewife syndrome, has dubbed it the 'problem that has no name'" (p. 16).

6. See Ehrenreich (1983) for a critique of these relatively "pro-feminist" pieces. Here, Ehrenreich challenges many of these pro-feminist pieces, asserting that they employ a problematic facsimile of liberal feminism that often works in favor of traditional masculinity, even as it purports a feminist position.

7. Indeed, Farrell's books appear in both pro-feminist and anti-feminist camps, indicating an apparent change in his perspective over the past few decades. If we could insert a category somewhere between these two camps, then Farrell's other work, *Why Men Are the Way They Are* (1986), would likely have to fit there.

8. For a discussion of public discourse around gay liberation, for instance, see Darsey (1994, 1981a, 1981b).

9. Illustrating these pressures, the article argues: "It is a stage of the election where the risks and the rewards are great for a challenger like Clinton. He told a crowd Saturday night that he sometimes felt bewildered by how far his campaign has come, a standard political gesture of humility but one with some credibility given Mr. Clinton's experience with near political death in the primaries. The atmospherics on the trail these days underscore the stakes: the heckling is tougher, the cheers more impassioned. A group of Bush-Quayle supporters, fresh-faced young people who said they were from local colleges, greeted Mr. Clinton at an event in a leafy suburb in Virginia on Saturday night. 'Hey, hey, ho, ho, draft dodgers got to go!' they chanted as his car rolled past the placards asserting, 'Clinton You're No Patriot.'"

10. "Who says Clinton is leaving no legacy? History will record that it was on his watch that the Food and Drug Administration approved the male impotence pill, Viagra, thus affirming the democratic principle that any man can aspire to be President" (p. 13).

Chapter 2

Little Big Men and Softhearted Hard Bodies: Homophobia as Hyper- and Hypomasculinity

One of the chief contradictions borne out in the Starr Report is the conflict between President Clinton's various degrees of sensitivity. This report casts Clinton as both an insensitive clod who leads on young Monica Lewinsky, apparently without any regard for her feelings, and a hypersensitive, hyperemotional wreck, whose emotional instability is itself evidence of the need for impeachment. Framing Clinton in this way, the Starr Report not only creates a paradoxical view of Clinton's sensitivity but also relates this confused hyper/hypo emotionalism to the American social body more generally. Here, Clinton's conflicted, on-again, off-again, sensitivity is constructed as evidence of his inability to lead the nation. President Clinton is either a callous, unfeeling tyrant who does not care about the laws of the nation or the American people (indeed, he is willing to lie to them), let alone Monica Lewinsky, or a bewildered lost soul who can barely handle his day-to-day phone calls, let alone run a nation.

While the Starr Report makes no mention of a crisis of masculinity, its rhetoric seems clearly to draw upon and extend these '90s discourses of masculine crisis, drawing on the sorts of conflicted notions of masculinity surrounding Clinton from his earliest presidential campaign onward. Variously praised for being a sensitive, sax-playing man-of-the-nineties, as well as for taking a hard stance on drugs, crime, and so on, even Clinton's positive evaluations illustrate the conflicted sensitivity possible for a '90s president. In a decade of increasingly vocal challenges to white male hegemony, such characteristics as sensitivity become more and more problematic. On the one hand, in the midst of challenges to traditional masculine anti-emotionalism, masculine sensitivity seems ever more cost-effective.

For advertisers, filmmakers, television producers, and even presidential candidates, shedding the trappings of the stoic, hardened macho man of the past for the heart of the sensitive "new male" seems a way to draw larger audiences by tapping into this desire for a transformed masculinity.

At the same time, this cost-effective "new man" must answer to a long history of representations stressing masculine toughness. Clearly, the stoic tough-guy images of the Marlboro Man or John Wayne are staple characters within this masculine history. Here, masculinity is figured as a storehouse of toughened warrior energy ready to be unleashed on any transgressive character. By nature of his contrast to this tough male past, the sensitive new male seems soft, passive, and weak. Challenging this sensitive new man, Robert Bly, for instance, argues in favor of this tough, warrior masculinity. "One man, a kind of incarnation of certain spiritual attitudes of the sixties, a man who had actually lived in a tree for a year outside Santa Cruz, found himself unable to extend his arm when it held a sword. He had learned so well not to hurt anyone that he couldn't lift the steel, even to catch the light of the sun on it" (Bly, 1990, p. 4). Here, "sensitivity" and "softness," the supposedly celebrated qualities of the new male, are seen as dangerous distortions of a man's tough, warrior character, leading to weakness, passivity, and impotence.

This seeming conflict between the new male's sensitivity and the toughness of the male of the past is hardly lost on savvy advertisers, producers, and directors. Addressing these media producers, John Fiske (1991) has argued that "in order to be popular, television must reach a wide diversity of audiences, and, to be chosen by them, must be an open text" (p. 347). Indeed, media producers of the '90s, rather than fully embracing either the new, sensitive male or the old, tough male often work to balance these seemingly conflicted images, working to maintain and extend their audiences as much as possible. Humorously contrasting "new men" with "old men," as in the sitcom *Frasier*, in which the wine-and-cheese-eating, Harvard-educated psychiatrists Frasier and Niles Crane live with their beer-swilling, retired police officer father; placing "old men" in new situations, as in the film *Junior*, in which former *Conan the Barbarian* Arnold Schwarzenegger becomes pregnant; rescuing old men from their old masculinity, as in *Regarding Henry*, in which a gunshot wound to the head transforms Harrison Ford from an aggressive mean-spirited attorney to a loving, sensitive father, while nonetheless keeping Henry's "toughness" in plain view; producers find a variety of ways to balance this supposed conflict between sensitivity and toughness.

Whereas Fiske's textual openness, or polysemy, stresses the possibility of multiple meanings and a tendency towards textual ambiguity, however, the sort of textual play illustrated in these '90s depictions of masculinity is of a different sort. Rather than simply leaving the movie or television program open for multiple readings, these media producers carefully manipulate this conflicted sensitivity, working to embrace the new male and yet maintain the importance of the old male as well. By holding this new masculinity in conversation with notions of masculinity from the past, these producers play both sides of this conflicted sensitivity, working to present products in tune with a new '90s masculinity, yet strong enough for the man of the past. This supposed conflict between the new, "sensitive" man and the old, "tough" man is the sort reflected in the Starr Report's characterization of President Clinton as well, illustrating the varied ways in which this conflicted sensitivity can be deployed. While media producers often balance this conflicted sensitivity to create "positive" characters, simultaneously "sensitive" and "strong," the Starr Report works to cast Clinton negatively as both sides of this sensitivity paradox, as both insensitive and too sensitive. Here, Kenneth Starr deploys this same conflicted sensitivity in an attempt to discredit Clinton as president and man.

To put this yet another way, these media portrayals of "new men," as well as the Starr Report's discussion of Clinton, demonstrate both the malleability and stability of sensitivity vis-à-vis masculinity. In an era dominated by a rhetoric of masculine crisis, "increased sensitivity" becomes, on the one hand, a convenient way to symbolize a move towards a new, more progressive man in touch with himself as well as with contemporary criticisms of masculinity. In this way, a film like *Jingle All the Way*, which has Arnold Schwarzenegger using his ex-ass-kicking brawn to find a Christmas present for his young son, may be seen to signal the arrival of a newly sensitive man, a movement away from the machismo of the past. Drawing upon this same rhetoric of masculine crisis, Kenneth Starr deploys sensitivity to figure Clinton as a hypersensitive jerk. That the Starr Report can frame Clinton simultaneously as hyper- and hyposensitive is indicative of the precarious state of sensitivity amidst the crisis of masculinity. Drawing on both more contemporary challenges to masculine anti-emotionalism and more traditional celebrations of hypermasculine toughness, the Starr Report's "sensitivity" is a paradoxical trait to be both celebrated and done away with.

This paradoxical sensitivity is the focus of this chapter. Both the Clinton of the Starr Report (as well as a host of other popular documents) and the "new man" of many contemporary media portrayals work to balance and exploit "new" and "old" notions of sensitivity. Exploring the conflict between these notions of sensitivity, as well as how they are negotiated within particular instances, allows us to diagnose and understand a set of anxieties at work within these various deployments. Specifically, this chapter interrogates two particular conceptions of masculinity, illustrating how each negotiates a '90s climate of sensitivity. First, I discuss the "hypermasculine hero," demonstrating how particular '90s actions heroes maintain a balance between hyper- and hyposensitivity, presented as "new men" even as they hold onto their traditional hypermasculine toughness. Steven Seagal, Patrick Swayze, and Wesley Snipes, the focus of this section, are fitting models of this odd balance of sensitivity. Second, I explore the "little man," another standard archetype of masculinity. With James Thurber's Walter Mitty as a standard archetype, I discuss the ways in which this character is depicted in contemporary television sitcoms. While the traditional, standard characteristics are present in these new portrayals, I point towards the highly sexualized nature of the '90s little man, a major shift from the little man of the past. Here, sensitivity is again oddly balanced, the '90s little man an interesting combination of hypersensitivity and hypersexuality.

Each of these deployments seems to demonstrate a move to accommodate a new, more sensitive masculinity of the '90s while at the same time upholding a traditional masculinity stressing toughness and machismo. While there is no necessary conflict between these various notions—between sexuality and sensitivity, for instance—these particular movies and television programs nonetheless figure a tension between "new" and "old" masculinities that demonstrates a sense of masculine crisis characteristic of the '90s. Likewise, through each of these deployments can be read an anxiety similarly characteristic of this '90s masculine crisis. The '90s little man, for instance, works to frame his sensitivity as hyper-*hetero*sexual, attempting to forestall any questions of the little man's sexuality. Such representations shed light on the workings of dominant notions of masculinity and their ritualization of particular anxieties, ideas important in understanding not only the media portrayals explored here but also the sort of masculine depiction present within the Starr Report. The conflicted sensitivity of these various depictions draws upon and contributes to a particular sense of masculine crisis, one entrenched within anxieties such as homophobia and tied firmly to traditional notions of machismo.

Having Your Machismo and Beating It, Too

In *Hard Bodies: Hollywood Masculinity in the Reagan Era*, Susan Jeffords (1994) discusses a 1980s "hard body" masculinity that employs "muscular physiques, violent actions, and individual determination" (p. 21), characteristics she links to the "Reagan Revolution" and, specifically, to Ronald Reagan's own masculine image. Evidenced by such blockbuster films as *First Blood* (1982), *Die Hard* (1988), *Lethal Weapon* (1987), *Terminator* (1984), *Robocop* (1987), *Raiders of the Lost Ark* (1981), *Top Gun* (1986), and *Batman* (1989) (pp. 16–17), this particular brand of Hollywood hypermasculinity offers American audiences "spectacular narratives about characters who stand for individualism, liberty, militarism, and a mythic heroism" (p. 16). Through a variety of these 1980s films, hypermasculine characters such as John Rambo, hard-bodied hero of *First Blood*, *Rambo: First Blood Part II* (1985), and *Rambo III* (1988), repeatedly strut their well-developed, hardened machismo, flexing their ever larger muscles and touting an ideal of masculine virility.[1]

At the same time, Jeffords also discusses a developing trend away from the hardest of these hard-bodied images. Reagan-era masculinity was constructed, somewhat paradoxically, as a combination of "the hard body and the 'sensitive family man,'" the Reagan Revolution "comprising on the one hand a strong militaristic foreign-policy position and on the other hand a domestic regime of an economy and a set of social values dependent on the centrality of fatherhood" (p. 13). Hence, Jeffords argues, the Reagan-era man is idealized as both muscular dynamo and sensitive father figure, seemingly blending the "sensitivity" of Ward Cleaver with the muscularity of John Rambo. As the '80s progress, Jeffords continues, this masculine duality increasingly moves towards the "family man" conception and away from the muscle-bound hypermasculine heroes of *Top Gun* and *First Blood*. Focusing on films such as *Kindergarten Cop* (1990), for instance, a film that domesticates Arnold Schwarzenegger from techno-muscular terminator to a "good with children" kindergarten teacher, Jeffords argues that "in 1991, the hard bodies of the 1980s seemed to have been successfully rejected in mainstream Hollywood films" (p. 140). Encouraged to move to a more domestic realm, the hard-bodied, hypermasculine hero experiences a sort of changed identity, seemingly challenging its traditional role as muscular-powered warrior and possessor of the hardest bodied machismo of the 1980s.

While these hard-bodied heroes seem to become "new men," their "old masculinity" never strays far from view, maintaining the sort of balanced, conflicted sensitivity mentioned above. Refusing to completely abandon this traditional masculinity, hypermasculine figures such as Steven Seagal, martial arts master of such action films as *Out for Justice* (1991), *Under Siege* (1992), *On Deadly Ground* (1994), and *Under Siege II: Dark Territory* (1995), embody "tough" machismo at its fullest, Seagal's precision bone breaking exemplary of the stereotypical macho male. Likewise, Wesley Snipes, manly hero (and sometimes villain) of such varied films as *New Jack City* (1991), *White Men Can't Jump* (1992), *Passenger 57* (1992), *Rising Sun* (1993), *Demolition Man* (1993), *Drop Zone* (1994) and *Money Train* (1995), holds a diverse range of filmic roles, each illustrating his hard-bodied machismo in a variety of ways. Finally, Patrick Swayze, whose similarly manly roles include a wide variety of films, from *The Outsiders* (1983), to *Dirty Dancing* (1987), to *Road House* (1989), to *Next of Kin* (1989), to *Ghost* (1990), to *Point Break* (1991), provides another interesting example for exploring '80s and '90s masculinity, his characters often dangerously destructive on the one hand while keenly sensitive on the other.

Increasingly new age, environmentally friendly, and politically correct, the pony-tailed Steven Seagal balances his tough persona with a mystical sensitivity that seems in line with contemporary criticisms of masculinity. Interested in changing "the essence of a man," as we will see below, Seagal's films seek a new masculinity that moves beyond the hyperviolent, insensitive masculinity of the past, even as Seagal himself embodies these old characteristics. Likewise, through an intertextual web that includes an "anti-masculine" stage of cross-dressed "gender bending" (with *To Wong Foo, Thanks for Everything, Julie Newmar* (1995), Patrick Swayze and Wesley Snipes illustrate a similar sort of conflicted sensitivity. With this film in particular, Swayze and Snipes, like Seagal, embrace a sort of new male sensitivity, yet simultaneously keep their hypermasculine "old maleness" in view, ultimately building the same sort of bizarre balance demonstrated in Seagal's films and elsewhere.

This discussion explores the seeming evolution of the hypermasculine hero from an '80s hard body of traditional manhood to a so-called new male hero of the '90s. Looking at the sorts of conflicted sensitivity evident within Seagal, Snipes, and Swayze, I offer a reading of the ways in which this new masculinity is tempered by reference to a more traditionally macho masculinity. These new male heroes embrace their new sensitivity with caution, their cautiousness evidencing anxieties implicit within this era of new masculinity and the contemporary crisis of masculinity.

Martial Arts Master as Hypermasculine Hero

The typical martial arts master presents an interesting brand of masculinity well suited for the conflicted sensitivity of the '90s. On the one hand, this character is a deadly force capable of the most calculated and precise destruction. His mastery of martial arts realizes a masculine ideal that traditionally values strength, "[admiring] fighting virtues and often [endorsing violence]" (Rotundo, 1993, p. 225).[2] This martial arts master develops skills in fighting that take this male celebration of power to its logical extreme, embodying a deadly precision that proves a model of masculine aggression. On the other hand, the martial arts hero traditionally demonstrates a near spiritual perfection that sublimates his aggressive power. David Carridine's character Kwai Chang Caine, of the mystical martial arts western *Kung Fu* (1972–1975) offers a '70s television example of this spiritually pure warrior. Caine is exiled from China after he kills the Emperor's nephew, who first kills Caine's teacher, Master Po. Fleeing to America, Caine spends his time wandering the American West in search of his half-brother, Danny Caine. The soft-spoken Caine speaks in haiku-like sentences that recall his spiritual education. Likewise, frequent flashbacks to his educational moments with Master Po further highlight the mystical nature of his character. Master Po, who calls the then-youthful Caine "Grasshopper," teaches Caine to be in touch with nature as well as his own feelings (perhaps offering a subtle nod towards the sensitivity training of the programs contemporary '70s). In the same way, Pat Morita's character, Mr. Miyagi, of *The Karate Kid* (1984) fame, illustrates a similarly sensitive inner spirit. A karate master who nonetheless enjoys the delicate art of bonsai tree trimming, Mr. Miyagi's inner sensitivity seems to balance his violent capability. Simultaneously sensitive and deadly, these earlier martial arts heroes illustrate a sort of conflicted sensitivity not unlike that characteristic of many '90s heroes.

With such a history established, the martial arts hero fits well within the particularly conflicted masculine sensitivity of the '90s. Contemporary martial artist Steven Seagal illustrates this conflict in exemplary fashion.[3] His expertise in fighting, his tough-guy, gangster-like voice, and his affinity for one-liners, reminiscent of fellow tough guys Schwarzenegger ("I'll be back") and Eastwood ("Go ahead, make my day"), seem to attest to his macho attributes. At the same time, recalling the mysticism of Caine and Mr. Miyagi, Seagal maintains a sort of empathetic sensitivity. "I want to know how to not be afraid to cry," Seagal told a group in Minneapolis, Minnesota, in the late '90s (Tillotson, 1999, p. F9), indicating his desire to get in touch with his sensitive side. That Seagal practices aikido, a martial art based on uniting with and redirecting, rather than merely opposing, an opponent's energy, also serves to highlight his mystical sensitivity. Taking these features together, Seagal's characters draw on the same seemingly ancient spirituality illustrated by Caine and Mr. Miyagi. Constantly delivering Confucian-like proverbs and reciting ancient tomes, Seagal seems to demonstrate a sensitive, new masculine character in contrast to the harder bodies of the '80s. Alongside the tough New York–Italian character he so often plays, Seagal's Asian wisdom and spirituality repeatedly illustrate the conflicted sensitivity so prevalent in the '90s.

Looking at one of Seagal's hard-punching films, *On Deadly Ground*, illustrates this conflicted notion of sensitivity more clearly. In particular, a pool-hall scene early in the film offers telling examples of this interesting conflict between Seagal's toughness and sensitivity as well as the conflicted, disturbing ways in which Seagal performs his "new masculinity." In this scene, set in an Alaskan barroom, Seagal, a Native Alaskan sympathizer, finds himself in conflict with a group of nonnative oil workers who taunt an intoxicated native man.[4] "I got Tanto with his fish breath bustin' my balls," the head oil worker begins, initiating a barroom confrontation in which he taunts the native man in a traditionally schoolboy bully fashion. Following this up, the worker cautions the native "Hey, Hey, don't bust my balls—I'm warning you." By threatening the native in this manner, a hardly politically correct act that demonstrates the oil worker's hypermasculine insensitivity, the oil worker highlights his stereotypically male aggression. Likewise, his obsession with his testicles as evidenced in his reference to "having his balls busted" calls further attention to his hypermasculine attributes, aiding in his association with stereotypically traditional masculinity. As the scene continues, the native stumbles around the increasingly angry men as the oil worker and his friends hurl expletives at the man and eventually push him to the floor. These last few acts finally drawing Seagal's attention; the oil worker challenges Seagal in typical hypermasculine fashion: "Hey, Cupcake, what the fuck you think you're looking at?" Exercising his aggressive abilities and calling attention to his phallic attributes, the oil worker establishes a clearly traditional, hypermasculinity against which Seagal's seemingly newer masculinity can begin to stand out.

"Nothing much at all," Seagal answers the oil worker calmly, walking to a far barstool and apparently diffusing the situation. By ignoring this confrontation (one of obvious tension, evidenced by the suddenly silenced onlookers anticipating a violent fight), Seagal distances himself from the oil worker's brand of hypermasculine insensitivity. Similarly, confronted by the oil worker's feminizing slurs such as "pussy" and "fucking pansy"—challenges that would surely constitute "fightin' words" for John Wayne or other such manly heroes— Seagal remains nonaggressive, further distancing himself from this more traditional masculinity. Engaging in a conversation with an elderly friend, Seagal ignores the taunting oil workers, refusing to encourage their hypermasculine bantering. Whereas the oil worker is overly aggressive and insensitive as demonstrated by his treatment of the native man as well as his willingness to taunt Seagal through feminizing and homophobic slurs, Seagal's unwillingness to participate seems to stress his empathy and multicultural sensitivity. Here is a more politically correct hero for a new age, the film seems to attest.

Maintaining his conflicted status, however, Seagal does not remain nonaggressive forever. Rather, just as the characters discussed above maintain both the new male's supposed sensitivity and the toughness of the old male, so Seagal eventually unleashes his old maleness. Once Seagal's conversation is finished, the oil worker and his friends harass the native for a second time, throwing beer in his face and pushing him once again to the floor. Here, Seagal begins to illustrate his aggressive, old male capabilities. Once the oil worker finishes with the native, challenging the entire barroom ("Who we gonna play with now?"), Seagal is first to respond ("Wanna play with me?"). Accepting the oil worker's challenge, Seagal obligates himself to a hypermasculine display of aggression. The manner in which Seagal fights these men, however, itself illustrates interesting masculine conflicts intended to separate Seagal from the more hypermasculine oil workers even as he himself engages in hypermasculine aggression. For instance, Seagal repeatedly strikes these men to the groin, a perfectly acceptable martial arts technique though a fighting strategy stereotypically reserved for women in the larger culture. From subtle side kicks to well-placed knees, Seagal dispatches blow after blow, repeatedly flooring each and every charging oil worker. Disabling these men by other means as well, this strike to the groin nonetheless appears Seagal's favorite, no other attack quite as effective in evoking a unison groan from his onlookers. As if this antiphallic theme were too subtle, several charging men let out painful screams of "my balls," making sure that the audience doesn't confuse these Seagal's blows for kicks to the stomach. When a scrappy oil worker grabs Seagal by the neck, he reaches back calmly, grabs the man between the legs and squeezes. The telltale crunching noise and the oil worker's painful scream—"My nuts!"—highlight the impact and direction of Seagal's attack. Seagal's hypermasculine aggression and his repeated strikes to these men's groins create a bizarrely interesting and conflicted image for Seagal's masculinity.

Nowhere, however, is this conflicted masculinity more obvious than in the final confrontation between Seagal and the head oil worker. Once Seagal has easily defeated the others, he and this merely old masculine figure face off in this ultimate showdown:

"Hey, you're a man, right? Are you a man?" Seagal begins, himself italicizing the oil worker's own traditional masculinity.

"Am I a man?" the oil worker echoes Seagal laughingly, moving to confront him face to face. "Yeah, I'm a man. I got a big pair of balls right between my legs." Through this assertion of manhood, and an accompanying motion towards the alleged "big pair of balls," the oil worker stresses still further his embodiment of an older, more traditional masculinity. Referring to his "balls" as evidence of his manhood, this oil worker illustrates a notion of masculinity heavily tied to aggression and virility. His is the conception of masculinity that urges men to "have some balls," in which one's balls serve as metaphors for traditional masculine toughness and the rejection of feminine emotion. Clearly, the movie suggests, this oil worker embodies a phallicentric conception of traditional masculinity of the sort critiqued within the '90s discussions of masculine crisis.

"Did you use those to beat up on this little native man?" Seagal asks, implicitly questioning this notion of a phallic, virile masculinity and suggesting his own, new male characteristics. In his following comments, Seagal satirizes this obsession still further, his own comments such as "big balls, okay," "you're a tough guy, got big balls," "here we go Mr. Big Balls," and "you're a man remember ... big balls between your legs?" serving parodic restatements of this phallic fixation. Through such remarks Seagal appears to challenge the traditional hypersexual, hyperaggressive, hyposensitive notions of masculinity that encourage men to "have some balls," defining masculinity in this metaphorically powerful way. Of course, to fulfill this phallic rejection Seagal must ultimately strike a blow to the oil worker's illusion of power. Thus, Seagal ends this standoff with a series of rapid strikes, several of which seem to hit their specific hypermasculine target. Injuring the oil worker and his self-acclaimed source of masculine power, Seagal leaves him a bloody, vomiting heap on the floor, Seagal's own hypermasculine aggression sufficient to defeat the aggressive tendencies of this macho oil worker—demonstrating both Seagal's ability to kick ass and his desire to separate himself from an older, ass-kickin' masculinity.

Yet, standing over this drooling, defeated man, Seagal apparently changes modes, illustrating once again his mystical sensitivity and empathy. For example, asking the oil worker gently, "What does it take? What does it take to change the essence of a man?" Seagal's tone changes from hypermasculine warrior to compassionate soothsayer. Converting back to this mystic, nonaggressive tone, Seagal returns to the antiviolent state evidenced in the scene's opening. Addressing the man in this empathetic, inquiring manner, Seagal contrasts himself again to the purely hypermasculine, old maleness of the oil worker. When the oil worker responds softly, "I need time to change ... time," Seagal's reply, "I do too, I do too," demonstrates further this empathetic, caring component of his character. Balancing hyper- and antimasculine characteristics in this manner, Seagal demonstrates his powerful, new sensitivity even as he flexes his powerful ass-kicking brawn.

Seagal's Anxious Sexuality (and Race and Class)

Seagal's final admission that he, too, needs time to change is an important moment for his conflicted sensitivity. On the one hand, Seagal's hardly subtle critique of traditional masculinity, illustrated in his mocking restatements of the main oil worker's claim, "Yeah I'm a man—I got a big pair of balls between my legs," clearly positions Seagal against this brand of hypermasculinity. Here, Seagal seems a sort of warrior against phallicentrism, going so far as to strike these oil workers in their alleged "big pairs of balls." Seagal seemingly wants to make it clear that he does not agree with the equation "masculinity equals big pair of balls." Of course, as already indicated, this rejection of traditional masculinity is not complete, highlighting the anxiety implicit within Seagal's new male sensitivity. While Seagal can initially ignore the homophobic taunts of the oil workers, eventually his traditional toughness must rear its head, rejecting the possibility of Seagal's homosexuality. Here, Seagal literally demands the phallus of heterosexual masculinity, his blows to the oil workers' "big pairs of balls," reasserting his own traditional toughness.

The other ways in which Seagal is "sensitivized" also evoke anxieties of traditional white masculinity. For instance, clothed as a Native Alaskan, Seagal seems a better native than the drunken man whom he rescues from the oil workers. Given the native man's drunken condition—evidenced through his slurred speech, motley dress, and staggering gait—Seagal is the only "native" who can legitimately care about the Alaskan environment or the pollution of the oil workers. Here, the "savagery" of the native, a traditional source of anxiety within white masculinity, is placated through another stereotype, that of the "drunken Indian," freeing native mysticism and spirituality for Seagal's co-optation. Similarly, the ways in which Seagal takes up an Eastern mysticism, via both his martial arts practice and the Eastern spiritualism repeatedly referenced within his films, evidence the incorporation of yet another culture, Seagal's sensitivity built through the colonization of one culture after another.

In this way, Seagal's sensitivity is conflicted in several respects. On the one hand, Seagal's "ass kicking" must stand as evidence of his traditional masculine toughness, maintaining the thoroughly hetero-sexual nature of his sensitivity. At the same time, Seagal demon-strates a similarly conflicted incorporation of these other ethnic and cultural groups, his Native Alaskan status, for instance, won only through discrediting and displacing that same culture. Finally, Sea-gal's relationship to the oil workers, whose job symbolizes the pollu-tion of Seagal's proclaimed native land, works to both incorporate and reject elements of working-class masculinity, demonstrating yet another mainstream masculine anxiety. Because working-class mas-culinity is positioned precariously between the working-class tough-ness and heroics of Rocky Balboa and the obnoxious, overweight likes of Archie Bunker, Seagal carefully negotiates these working-class stereotypes. While Seagal is generally presented as a working-class hero, whether as a rugged New York cop or as a Texas cowboy, the stereotyped portrayal of working-class men like the oil workers works to distance Seagal from the Archie Bunkeresque characteristics represented within mainstream media culture—embracing working-classness at the same time that it is held at bay. In these ways, Sea-gal's sensitivity is won via the colonization of native and working-class cultures (taken up more fully in chapters to follow), as well as the assertion of a homophobic heterosexual toughness, contributing to a structure of feeling that maintains these particular anxieties throughout.

Cross Dressing and Hypermasculine Sensitivity

> [T]he transvestite states the question: "When I am like a female, dressed in her clothes and appearing to be like her, have I nonetheless escaped the danger? Am I still male, or did the women succeed in ruining me?" And the perversion—with its exposed thighs, ladies underwear, and coyly covered crotch—answers, "No. You are still intact. You are male. No matter how many feminine clothes you put on, you did not lose that ultimate insignia of your maleness, your penis." And the transvestite gets excited. What can be more reassuringly penile than a full and hearty erection? (Garber, 1992, p. 94)

Similarly to Seagal, Wesley Snipes and Patrick Swayze offer their own negotiations of this conflicted '90s sensitivity. Whether ripping out a villain's throat, as Swayze (Dalton) does in *Road House* (1989), or playing a macho villain trading blows with Sylvester Stallone, as Snipes (Simon Phoenix) does in *Demolition Man* (1993), these two masculine icons embody the same sort of hypermasculine aggression evidenced by Steven Seagal. Still, Swayze and Snipes both portray a wide range of filmic characters. Snipes' characters, for example, include "Web Smith" in *Rising Sun* (1993), Snipes playing a savvy detective partnered to Sean Connery, as well as "Sidney Deane" in *White Men Can't Jump* (1992), Snipes here playing a talented basketball hustler. Likewise, Swayze portrays such varied roles as "Bodhi" in *Point Break* (1991), playing the skydiving antagonist to hero Keanu Reeves, and "Sam Wheat" in *Ghost* (1990), the sensitive '90s man who comes back from the dead to solve his own murder and visit wife, Demi Moore. Each of these films paints these characters' masculinity in a slightly different manner, providing a sense of depth perhaps missing from Seagal, yet still maintaining both their new male sensitivity and their old male hyperaggressive toughness.

   None of these roles, however, does more to demonstrate Swayze and Snipe's conflicted masculinity than their 1995 roles in *To Wong Foo*. Here, with their young drag queen protégé Chi-Chi Rodriguez, Noxeema (Snipes) and Vida (Swayze) drive west from New York to Hollywood, spreading their cross-dressed charm wherever they stop. Dressed in heels, their lips carefully painted with lipstick, these action-heroes-turned-drag-queens taunt their stereotypically masculine personalities, seemingly subverting their masculinity through this new code of dress and conduct. Yet, their "harder bodied" masculine characteristics always close at hand, Snipes and Swayze seem to make a case for their toughness throughout, illustrating the same conflicted, anxious sensitivity maintained by Seagal.

   The cross-dressing theme exploited in *To Wong Foo* follows an interesting history that makes it particularly appropriate for this partial subversion of traditional masculinity. A well-established theme by 1995, this transvestitism can eventually be absorbed by hypermasculine heroes like Snipes and Swayze, bringing it within their repertoires of their negotiated sensitivity. Depicted in films such as *Some Like It Hot* (1959); *The Rocky Horror Picture Show* (1975), a continuing cult classic that often has audience members themselves dressed in drag; *Tootsie* (1982), with Dustin Hoffman playing a cross-dressing actor; *The Crying Game* (1992), the '90s film that had mainstream audiences "conspiring" to keep the secret of the "leading women's" penis;[5] *Mrs. Doubtfire* (1993), with Robin Williams as a husband turned cross-dressed nanny; *The Adventures of Priscilla Queen of the Desert* (1994), an Australian counterpart to *To Wong Foo*; and Rupaul's *Wig-stock* (1994); as well as in television shows such as *Bosom Buddies* (1980–1982), this cross-dressing iconography develops over time, proving a notion ready-made for this conflicted '90s deployment.

Dustin Hoffman's *Tootsie* character, Michael Dorsay (Dorothy Michaels), who dons a wig and dress to land a female lead in a soap opera, helped this cross-dressing theme gain a continuing foothold in mainstream culture, showing a "non-hard-bodied" man (at least relative to fellow '80s men Schwarzenegger and Stallone) who seemingly adopts transvestitism out of economic necessity. Similarly, Robin Williams' Daniel Hillard (Mrs. Doubtfire) wears this cross-dressed garb out of a like necessity, posing as a nanny in order to spend time with his children once he and his wife have divorced. These various films work to partially desensitize audiences to this cross-dressing film phenomenon, getting them used to the filmic sight of men in drag. Likewise, the desexualized necessities that force Williams (Hillard) and Hoffman (Dorsay) into this feminine dress work to counteract the more overtly sexual depictions of *The Rocky Horror Picture Show* and *The Crying Game*, making possible a mainstream, filmic transvestitism that blunts the subversive edge and sexual tone of these gender-bending images. Here, cross-dressing becomes a purely comedic vehicle that lacks much of its otherwise subversive, threatening capacity.

In the same way, *Martin* sit-com character "Sheneneh," played by a cross-dressed Martin Lawrence, illustrates another contemporary image of male cross-dressing. A campy version of this gender bending, Sheneneh's melodramatic antics, as well as her comedically accentuated figure, constantly remind the viewer that this is Martin Lawrence playing at playing female. Likewise, *In Living Color's* comedic, man-in-drag character "Wanda" presents a similarly satiric look at cross-dressing, her bizarrely proportioned body, and her overused one-liner, "I'm gonna rock your world," powerful enough to scare away any of her would-be suitors. Here, both of these '90s television shows parody this idea of cross-dressing, further softening its subversive possibilities and making it more acceptable as a mainstream vehicle.[6] Placing these various "non-hard-bodied" men in drag, from Dustin Hoffman to Tom Hanks (*Bosom Buddies*) to Robin Williams, films and television programs of the '80s and '90s create a mainstream transvestitism that becomes more and more available for "harder bodied figures." Thus, this cross-dressing is broken in by the non-hard-bodied likes of Williams, Hoffman, and Hanks, as well as the campy satire of Martin Lawrence, eventually creating an avenue through which even hard-bodied stars Snipes and Swayze may embrace this gender-bending without completely compromising their hard-bodied personae.

Continuing this idea, Marjorie Garber (1992) discusses the ability of transvestitism to both subvert *and* maintain traditional notions of masculinity. Quoting Dr. Robert Stoller, a psychoanalyst who has written extensively on issues of cross-dressing and transvestitism, Garber writes:

> The whole complex psychological system that we call transvestism is a rather efficient method of handling very strong feminine identifications without the patient having to succumb to the feeling that his sense of masculinity is being submerged by feminine wishes. The transvestite fights this battle against being destroyed by his feminine desires, first by alternating his masculinity with the feminine behavior, and thus reassuring himself that it isn't permanent; and second, by being always aware even at the height of the feminine behavior—when he is fully dressed in women's clothes—that he has the absolute insignia of maleness, a penis. And there is no more acute awareness of its presence than when he is reassuringly experiencing it with an erection. (Garber, 1992, p. 95)

At once both "antimasculine" and "masculine affirming," this conventionalized understanding of cross-dressing offers an inviting addition to the hard body's negotiation of old and new maleness. Coupled with the already developed, complex, masculinity of Swayze and Snipes, for instance, this complex, blunted cross-dressing maintains these characters' conflicted hypermasculine heroics, neither fully rejecting nor embracing their powerful machismo.

Swayze's own filmic masculinity has a long history, including such films as *The Outsiders* (1983), *Dirty Dancing* (1987), and *Next of Kin* (1989), not to mention *Road House*, *Ghost*, and *Point Break*. Similarly to Seagal, Swayze's characters such as Dalton and Bodhi demonstrate a competence in the martial arts; Dalton, for instance, is a barroom bouncer who readily dispatches the most disruptive of customers, calmly escorting them to the door with holds and joint locks that would make Seagal proud. Likewise, his earlier character Darrel Curtis, in *The Outsiders*, is a tough greaser who also serves as a father figure for younger brother Pony Boy. Even dance instructor Johnny

Castle, Swayze's character in *Dirty Dancing,* is portrayed as a working-class "tough guy," a hard-core rebel who punches a hole in his car's window when he locks the keys inside. At the same time, Swayze's characters also display a level of romanticized sensitivity that complements his hard-bodied machismo. Dalton, for instance, is a student of Eastern philosophy. Johnny Castle's fluid ballroom dance stylings demonstrate a grace and romantic charisma that makes his female dance students swoon. *Ghost's* Sam Wheat agonizes through a purgatorial quest to unite with his wife, the now infamous "potter's wheel" sex scenes evidence of his powerful passion. These romantic characters and characteristics, along with his hard-bodied tough-guy persona, make Swayze the ideally macho hero that he is. Successful in love and war, Swayze seems both tough and sensitive, a seemingly perfect negotiation of old and new conceptions of maleness.

Similarly to Swayze, Snipes offers a near picture-perfect figure of conflicted masculinity. In *White Men Can't Jump,* Snipes' trash-talking, basketball-hustling Sidney Deane makes short work of opponents, his verbal cuts nearly as powerful as his slam dunks in chastising and defeating his unfortunate victims. Likewise, Snipes' characters continuously demonstrate their competence in martial arts, evidencing an aggressive power not unlike Swayze and Seagal. In a typical Snipes' scene from *Drop Zone,* for instance, his character, Pete Nissip, bursts into a restroom, aware that a gang of skydiving thugs is beating one of his allies. Without overstraining his hypermasculine muscles, Nissip easily defeats them, blow after powerful blow always hitting his specific target. Through these characters, like Swayze and Seagal, Snipes portrays a powerfully masculine hero, a hero ripe for this '90s negotiation.

In stark contrast to these hypermasculine images, Noxeema Jackson and Vida Boheme work to counter Snipes' and Swayze's traditional characters. Recognizing this, in a CNN interview, Snipes explained his choice to play Noxeema, stating "I don't want to be associated with the martial arts action personality; that's what I found myself in, but I don't even think that's my strong point." Likewise, Swayze called his role as Vida a "challenge, that's what it was, could I pull it off … and touch people's hearts with a character that was so far from me." Indeed, as Vida Boheme, Swayze plays against both of his macho images, contradicting his hard-bodied martial arts persona as well as his image as smoldering male sex symbol. Donning "her" dress and wig, Vida seems to scoff at traditional masculinity, telling listeners to "believe in yourself, and moisturize—this I cannot stress enough," and projecting an attitude seemingly far from Swayze's own hypermasculine stereotype. Whether sashaying down the runway in a drag queen beauty contest, or ordering her favorite drink, a "pousse cafe," Vida Boheme appears a powerful challenge to Swayze's masculinity, undercutting his hard-bodied image.

Likewise, Snipes' Noxeema Jackson struts her cross-dressed persona, similarly challenging Snipes' hard-bodied machismo. At a club, for instance, real-life supermodel Naomi Campbell tells Noxeema, "I wish I was as beautiful as you," reiterating the glory of Noxeema's "femininity." Similarly, when the "girls" stumble onto an attic full of vintage clothing in Snydersville, the hick town they transform with their transvestite magic, Noxeema squeals in celebration "Ooooooh! Vida! It's all from 1966! Fringe! I'm gonna black out! Clara Pearl, we gonna make you look like Emma Peel! There's gonna be a summer of love on our bodies." Further demonstrating this fashion-sensed excitement, she comments "What about that brown thing over there? Would I wear it the whole day? I can groove on that. Any boots?" Maintaining this keen sense of fashion, Noxeema proves her right to the title of drag queen. After all, she explains, "When a gay man has way too much fashion sense for one gender, he is a drag queen." Hardly Pete Nissip or Sidney Deane, Snipe's Noxeema seems to counter Snipes' traditionally masculine character, offering instead a fashion-conscious cross-dresser who understands the importance of makeup and drinks his "Gibson, straight up with a twist."

Ultimately, however, Snipes' and Swayze's respective masculinity shines through, providing a paradoxical picture that, as with Seagal, works to maintain their old masculine toughness throughout. Commenting on this idea, movie critic Sean P. Means (1995) of *The Salt Lake Tribune* observes, "with Swayze and especially Snipes—whose bulging biceps explode any feminine illusion—you never forget you're watching macho men playing at drag." Likewise, throughout the film, various incidents and scenes seem to call out to Snipes and Swayze's traditional, hard-bodied masculinity, keeping it readily available. For instance, when the "girls" stop at a motel, relieved to find a women's basketball convention in progress (a scene these oversized, cross-dressed "women" fit into well), Noxeema suddenly turns into Sydney Deane, stealing the court as she dunks over the heads of her basketball-playing sisters. Further, when Snydersville's young men get rude with the local women, Noxeema executes a "Seagal-esque" groin grab, using her enormous strength to force an apology and encourage these hooligans to act more gentlemanly. Addition-ally, as Garber (1992) suggests, Noxeema's (Snipes') gender-bending dialectic is complicated still further because, "paradoxically, the black American man has been constructed by majority culture as both sexually threatening and feminized, as both super-potent and impo-tent" (p. 271). Because Snipes' masculinity is already paradoxical, his cross-dressing serves to reinforce and amplify this contradictory mas-culinity, the opposing images of Snipes' action heroes and the cross-dressed Noxeema, providing layers of contradiction upon which to build this negotiated masculinity.[7]

Similarly, Swayze's Vida Boheme maintains this conflicted mascu-linity by portraying shades of the more hypermasculine figures of Swayze's earlier films. When a backwoods sheriff, Sheriff Dollard, gets fresh, believing he has just what this "career girl" needs, Vida surprises him, knocking him to the ground and leaving him for dead. Similarly, Vida apparently "turns Dalton" when a Snydersville hus-band gets abusive with his wife, letting the abuser know that she cannot tolerate such behavior. As if this were too subtle a message, in another Snydersville scene, when Vida, Nox, and Chi-Chi get in a fight, frustrated by the conditions of the backwoods town, Vida has her wig pulled off showing her own hair for the first time since the movie's opening. Looking at "her" reflection in the mirror, this bald Vida stands as a reminder that, beneath the dresses and makeup, the macho Swayze is, indeed, still present.

Mainstreaming the Transvestite

A number of critics observe that *To Wong Foo* seems to consciously avoid the topic of sexuality, apparently a subject too risky for this mainstream transvestitism. Anthony Lane (1995), for instance, argues that "one of the creepy, disingenuous aspects of *To Wong Foo* is that it uses drag as a convenient way of not thinking about sex. From time to time, Noxeema and Vida drop the word 'gay' into the conversation, but that is the sole sign of anything stirring in the hormone depart-ment" (p. 96). To the end, *To Wong Foo* sticks to its mainstream trans-vestitism, a model culled and developed from films like *Tootsie* and

Mrs. Doubtfire, that "sensitivizes" Swayze and Snipes, while holding to their traditional macho images.

This particular mainstream transvestitism functions similarly to Steven Seagal's presumed critique of "big-balled masculinity" illustrated above. Just as Seagal eventually shows his tougher side, proving his traditional heterosexual toughness, so Swayze's and Snipes' characters have their mainstream, traditionally masculine characteristics as an explicit subtext. By both reminding the viewer of their hypermasculine toughness and by desexualizing their characters (indeed, during the film only John Leguizamo's character, one without the hypermasculine tradition of Swayze and Snipes, shows feelings for another man), Swayze's and Snipes' characters embrace the "sensitivity" implied in their mainstream transvestitism while homophobically rejecting the possibility of an alternative sexuality. What might serve as a departure from their more traditionally hypermasculine roles—a "character so far from" this masculine tradition—instead works implicitly to strengthen that tradition.

In the same way that Seagal colonizes Native Alaskan culture in order to win his new male sensitivity, so Swayze's and Snipe's characters win their new '90s masculinity through a co-optation of drag, pacifying its subversive potential. While drag may serve to parody notions of masculinity and femininity, showing the fictive nature of these gender constructions (see Butler, 1990), in the hands of Swayze and Snipes, this parody is placated and, indeed, reversed, offering a parody of drag's parodic potential. Keeping Swayze's and Snipe's hypermasculine traditions close at hand, *To Wong Foo*'s drag becomes a performance that merely highlights the ridiculousness of hiding these masculine pasts. By posing Swayze and Snipes as drag queens, *To Wong Foo* feigns a comfort with gender bending that might indeed seem to dislodge the hypermasculine histories that have made these male heroes famous. Yet, by constantly clinging to these very histories, *To Wong Foo*'s mainstream transvestitism naturalizes and reiterates the heterosexist anxieties drag may otherwise expose and exploit. In the end, the movie assures us, no dress can hide Swayze's and Snipes' hypermasculinity.

Little Big Men

While 1990s Hollywood films like those of Swayze, Snipes, and Seagal, awkwardly attempted to embrace a so-called new maleness, others were slightly less subtle about their anxieties. Asserting that feminist discourse had caused problems for contemporary masculinity, Robert Bly (1990) argued that:

> As men began to examine women's history and women's sensibility, some men began to notice what was called their feminine side and pay attention to it. This process continues to this day, and I would say that most contemporary men are involved in it in some way. There's something wonderful about this development—I mean the practice of men welcoming their own "feminine" consciousness and nurturing it—this is important—and yet I have the sense that there is something wrong. The male in the past twenty years has become more thoughtful, more gentle. But by this process he has not become more free. He's a nice boy who pleases not only his mother but also the young woman he is living with. (p. 2)

Bly sees this soft maleness as widespread in his contemporary culture. "Sometimes even today when I look out at an audience, perhaps half of the young males are what I'd call soft," he continues. While Bly sees soft males as "valuable people—I like them—they're not interested in harming the earth or starting wars" (p. 2), he still sees a problem with this masculine incorporation of feminist ideas. Although getting in touch with one's feminine side may be useful in certain ways, this feminization made Bly nervous. Bly's reaction thus reiterates the argument made by numerous masculinity scholars that such potentially countermasculine discourses of feminism and sexuality can cause crises for contemporary men like Bly, as well as for their contemporary notions of masculinity, forcing them to reevaluate and renegotiate their senses of masculinity.

This next discussion explores such anxieties over sensitivity and soft masculinity by investigating the contemporary use of the classic, archetypal character of "the little man." A discussion of James Thurber's short story "The Secret Life of Walter Mitty" highlights the classic characteristics of this image. Walter Mitty, the story's main character, is controlled by the circumstances of his life, his authoritative boss, and his outspoken wife. He escapes these controls and finds some measure of autonomy only within the elaborate fantasies and daydreams he concocts for himself. While Mitty is powerless in his everyday life, he achieves a sense of power by dreaming himself into one powerful male after another. Similarly to Walter Mitty, contemporary sitcom characters Martin Tupper (Brian Benben) of television's *Dream On* and Ross Geller (David Schwimmer) of *Friends* are both similarly controlled by the circumstances of their lives. Divorced by his wife, Judith, and living in the shadow of her new husband, world-class surgeon Richard Stone (aka Dick Stone), Martin evidences the same sense of powerlessness that characterizes Walter Mitty. Likewise, Ross' wife has left him for a woman, he has to give his pet monkey Marcel to the zoo when he is unable to control him, and he spends the early seasons of the show pining after a woman to whom he can never express his true feelings, expressing his affection only later in the series. Like Mitty, both Ross and Martin appear helpless men struggling for autonomy within their contemporary worlds, but having only their dreams to keep them going.

In contrast to Walter Mitty, however, these contemporary little men are hypersexualized in ways that seem to contradict their littleness. Martin, for example, is extremely successful sexually, sleeping with a host of stereotypically attractive women, from supermodels to rock stars to young coeds half his age. Similarly, popular discussions of David Schwimmer's character, Ross, repeatedly make reference to his sexuality, stressing his attractiveness to women and "waxing poetic over his puppy dog eyes" (Kahn, 1996, p. 48). While Walter Mitty appears impotent, even asexual, contemporary little men Ross and Martin are presented as highly sexualized and sexual, despite their Mitty-like lack of power in the rest of their lives. Ultimately, the hypersexualization of these popular little men evidences the same sense of crisis discussed in the other popular discourses addressed above. Specifically, in a time marked by multiple challenges to manhood, from feminist critiques of patriarchy, to gay and lesbian deconstructions of heterosexuality, the little man, like maleness more generally, seems stuck in a precarious position. Just as Robert Bly both applauds and winces at his contemporary soft male, so the little man both celebrates and is uncomfortable with his own littleness.[8] In the end, similarly to more hypermasculine characters Seagal, Snipes, and Swayze, the little man's hypersexualization proves hyper-heterosexual, even homophobic, an attempt to both celebrate the sensitive male and to protect that male from discourses of gender and sexuality that comprise this contemporary crisis of masculinity.

Walter Mitty: The Impotent, the Pitiful, the Humiliated

"We're going through!" The Commander's voice was like thin ice breaking. He wore his full-dress uniform, with the heavily braided white cap pulled down rakishly over one cold gray eye. "We can't make it sir. It's spoiling for a hurricane, if you ask me." "I'm not asking you, Lieutenant Berg," said the Commander. "Throw on the power lights! Rev her up to 8,500! Were going through!"

So Walter Mitty envisions himself during the daydream that opens Thurber's short story, "The Secret Life of Walter Mitty."[9] Whether dreaming of himself as the commander of a battleship, a world-class surgeon, a sharpshooter on trial for murder, or a convicted criminal standing before the firing squad, Mitty constructs himself one heroic image after another. Mitty's dreams of power remain inevitably short lived, however, and are continually interrupted by the questions and commands of his wife, police officers, parking lot attendants, and even people passing him on the sidewalk. As much as Mitty may dream of worlds where he can live an adventuresome, exciting lifestyle, the mundaneness of his everyday life consistently squelches these big dreams. This opening dream, for instance, in which Mitty sees himself as the commander of a battleship fearlessly leading his crew through treacherous waters demonstrates the sort of world that Mitty hopes to create for himself and thus the world that he hopes to escape. Pressing on in the face of danger, barking orders to his crew, "Switch on No. 8 auxiliary!," "Switch on No. 8 auxiliary!," "Full strength in No. 3 turret!," "Full Strength to No. 3 turret!," the commander has a powerful persona, one recognized and respected by the ship's crew. "The Old Man'll get us through," the crew says to each other, grinning with confidence. "The Old Man ain't afraid of Hell!" Dreaming these personas, Mitty attempts to gain a sense of masculine power in his otherwise powerless life.

In contrast to the commander, Walter Mitty himself is afraid of hell, as well as the little hell on earth in which he feels himself constantly trapped. The story demonstrates this sense of containment and suffocation each time Mitty is pulled from his dream world by the interruptions of the people through whom he feels his power arrested. "Not so fast! You're driving too fast!" his wife cuts short his adventures as the commander. "What are you driving so fast for?" These repeated interruptions consistently emphasize Mitty's subservience to those around him, divesting him of the hyperauthoritative power he imagines within his dreams. "You were up to fifty-five," his wife continues. "You know I don't like to go more than forty." While the commander of whom Mitty dreams will push on through icebergs or dangerous waters, Mitty himself must maintain the speed limit dictated by his wife.

Mitty's dissatisfaction with the split between his ideal world and the "real" world to which he is constantly returned manifests itself in various ways throughout the story. For instance, when his wife reminds him to "remember to get those overshoes while I'm having my hair done," Mitty challenges her, "I don't need overshoes," attempting to assert himself in a way more in line with the commander. That his wife needs only comment "We've been all through that" to convince him to purchase the shoes, however, demonstrates the disjunction between the two worlds that Mitty attempts to negotiate. In another seemingly resistant act, Mitty puts on his gloves when his wife asks him to, but quickly removes them again once she has left the car. When a cop yells at him to "pick it up buddy," however, "Mitty hastily pulled on his gloves and lurched ahead," illustrating his acquiescence to his wife's original request.

It is telling that Mitty conflates the police officer's command, "pick it up buddy," with his wife's earlier request that he wear his gloves. Unable to separate one seemingly authoritative voice from another, Mitty's entire world seems positioned in control of him, pointing towards the powerlessness he feels and repeatedly demonstrates. Even the parking lot attendant at his wife's hair salon seems in more control than Mitty. "Back it up, Mac! Look out for the Buick!" the attendant yells, breaking Mitty from yet another dream of personal power. "Wrong lane, Mac," the attendant adds, with Mitty's quiet muttering "Gee. Yeh," highlighting his sense of helplessness. Similarly, when the attendant, who offers to park the car for Mitty, "vault[s] into the car, back[s] it up with insolent skill, and put[s] it where it belong[s]," Mitty's own inabilities are drastically foregrounded. "'They're so damn cocky,' [thinks] Mitty, walking along main street; 'They think they know everything.'"

The "they" that Mitty refers to here addresses not only the parking attendant himself but the rest of the world against which Mitty feels himself in constant struggle. This struggle plays itself out in the final scene of the story, in which Mitty, waiting for his wife, dreams himself a glorious, heroic death:

> Walter Mitty lighted a cigarette. It began to rain, rain with sleet in it. He stood up against the wall of the drugstore, smoking…. He put his shoulders back and his heels together. "To Hell with the handkerchief," said Walter Mitty scornfully. He took one last drag on his cigarette and snapped it away. Then, with that faint, fleeting smile playing about on his lips, he faced the firing squad; erect and motionless, proud and disdainful, Walter Mitty the Undefeated, inscrutable, to the last.

While this ends the story proper, were the story to continue, some voice from the real world would no doubt call Mitty back to the mundane existence he tries so hard to escape—just as it does from each of his other dreams. Hence, while Mitty may dream himself to be "erect" and "motionless," "proud" and "disdainful," "undefeated," and "inscrutable," his real-world self inevitably feels limp, humiliated, and defeated.

To the end, Mitty equates personal power with the hypermasculine fantasies he creates for himself. Impotent and powerless over his surroundings he escapes this hell via his elaborate dreams of masculine prowess and virility. The commander, the sharpshooter, the surgeon, and the convicted felon all evidence Mitty's image of hypermasculine strength. Mitty's dreams are constantly interrupted, highlighted as mere fantasy, evidencing his inability to achieve these supposed ideals. Ultimately, however, Mitty cannot even be called tragic, cannot say "look how much I've suffered." The so-called tragedies against which he struggles—buying shoes, wearing gloves, finding the right brand of puppy biscuits—seem out of proportion with the importance he places upon them. Ever the helpless little man, Mitty struggles in a world in which he feels utterly powerless.

As powerless and pitiful as he may be, however, Walter Mitty is not without his charm. According to Anthony Kaufman:

> The story has become folksy—Mitty is seen as endearing, the amiable little man, who dreams his dreams like all of us, and who triumphs in his dreams over the dull, gray world of suburban (and for that matter urban) America. The feature film made of the story in 1949 presents him thus; and two generations have seen him as a character rather like Dagwood Bumstead: The American middle-class everyman, presented to us with a wry but friendly smile. (Kaufman, 1994, p. 93)[10]

It is this folksy, endearing quality that opens Mitty, and the little man more generally, for contemporary rearticulations. That Mitty proves likable even while his masculinity seems in question makes him an interesting persona for the contemporary crisis of masculinity. Taking up these characteristics, characters such as Martin Tupper and Ross Geller can themselves offer negotiations of contemporary notions of masculinity. However, their hypersexualization, a stark contrast to Mitty's impotence, positions these contemporary little men in significantly different ways.

Martin and Ross: The Hypersexual Little Man

Sitcom characters Martin Tupper and Ross Geller both exemplify a number of the characteristics displayed by Walter Mitty. Martin is himself a sort of dreamer. Most obviously, the brief "daydreams" in the form of clips from classic films and television programs that constantly interrupt the diegesis of *Dream On* demonstrate Martin's subconscious desires and the thoughts he contemplates but does not articulate to the other characters within the program. For instance, when an editor comments favorably on one of Martin's manuscripts, the scene cuts to Lucille Ball patting herself on the back. When Martin's son Jeremy offers to loan Martin money from the trust fund set up by his stepfather, the scene flashes to a Western's image of a man getting shot in the chest. Time and again, Martin's inner thoughts are represented in these Mitty-like dreams. Like Walter Mitty, Martin dreams a world separate from the one in which he lives.

While Mitty imagines himself as the commander or fantasizes that he is a world-class surgeon, however, Martin's dreams are centered on the media, as illustrated in these film and television clips. The opening segment of *Dream On* likewise reiterates the importance of media within Martin's life. Depicting Martin as a child in front of the television set, this beginning sequence presents him as a media child raised on the consumption of these mediated images. That Martin's subconscious thoughts are illustrated via media clips demonstrates his even larger, Mitty-like dream, namely his desire to be a media star himself. Like Mitty, Martin is never able to achieve this dream, his most earnest attempts always somehow falling short. For instance, Martin works as a book editor and here dreams of the great books his mediocre manuscripts might become. A wannabe writer, he more than once tries to write the great American novel and is disappointed, at one point, when his teenage son manages to write a best-seller instead of him. He constantly fights with his boss, Gibby, as well, pushing him to publish more serious works of literature, apparently in hopes of achieving some vicarious greatness in this way. Similarly, when he tries his hand at playwriting he gains critical recognition only by writing pornographic films; his pornographic masterpiece, *Catch Her in the Rear*, is heralded by one critic as the *Citizen Kane* of pornography ("This is not what I would call a porn movie," the producer tells him. "This, my young friend, is porn cinema"). Each of these incidents rings with Mitty-like qualities; Martin constantly dreams of the "great man" he could become, but somehow always manages to fall short.

To make matters worse, Martin constantly lives in the shadow of his ex-wife's husband, billionaire and Noble Prize–winning genius Richard Stone (that is, Dick Stone). Richard's many accomplishments include, among countless others, running a shelter for the homeless, membership on the U.S. Olympic equestrian team, a Grammy nomination for an album of folk songs, and operating on Barbra Streisand's godchild. Stone rarely appears on the program, usually because he is busy accepting another award, performing a good deed, or generally living up to his reputation as a genius, do-gooder surgeon. The rare times that he does appear, however, it is as a beam of light, glowing in the doorway of Martin's apartment. Godlike in both activity and appearance, Richard Stone is the man Martin, like Mitty, would be, if only his dreams could come true.

Similarly to Martin, Ross Geller of *Friends* demonstrates his own Walter Mitty-like characteristics. Ross is a paleontologist and as such spends his time in childlike dreams of dinosaurs. He also spends the early episodes of the program dreaming of a relationship with Rachel Green (Jennifer Aniston), unable to voice his feelings for her until much later in the series. Rather, the oft tongue-tied Ross mumbles and stumbles his way through these early shows. Ever ready to express his true feelings, he ultimately ends up silenced by his lack of confidence and his fear that Rachel will reject him. Like Mitty and Martin, Ross keeps much of his dreaming to himself (though he often shares it with his other "friends"), his inability to express his feelings to Rachel evidencing this fairly clearly. When Ross does finally express his love for Rachel, he is unable to sustain their relationship in a meaningful manner. While this continuing tension serves as a generic convention, maintaining the humorously unfulfilled love conflicts that are a mainstay of television situation comedy, it also works to highlight Ross' powerlessness still further. For instance, Ross and Rachel eventually break up when Ross becomes jealous of Rachel's relationship with a studly male co-worker. Feeling powerless in the face of this other male, Ross eventually runs away from Rachel, anxious that he cannot compete with this more powerful man. Like Martin's relationship to his godlike nemesis, Richard Stone, so Ross feels little in this other man's shadow.

Still more clearly, like Mitty and Martin, Ross illustrates a powerful level of angst, suffering the powerlessness and failures as a husband and father that seem to characterize the little man. An article in *GQ* sums these problems up nicely:

> Ross has taken some tough knocks. He's a museum paleontologist whose pregnant wife divorced him for her lesbian partner. He's a baby-smitten new dad who has to listen to his son's gleeful "other mommy" describe the taste of birth mommy's breast milk ("kinda like cantaloupe juice"). The capuchin monkey Ross adopted to counter the loneliness had to be sent to a zoo, owing to an unseemly testosterone surge. (Hirshey, 1996, p. 234)

Indeed, Ross seems impotent as both husband and father. Just as Martin lives in the shadow of Richard Stone, Ross lives in the shadow of his wife's lesbian lover, his son's "other mommy." Similarly, that he cannot even control his pet monkey, who frequently plays songs over and over against Ross' expressed wishes, brings his lack of control into full view.

In particular, that Ross' wife has left him for her lesbian lover is a failed dream of the highest order. While lesbianism may indeed be a common male fantasy, in Ross' case this dream has backfired in the most profound of ways. Unable to enjoy the lesbian fantasy that Joey and Chandler, Ross' other male friends, find so exciting, Ross feels disempowered and displaced from his roles as father and lover. With this other mommy seemingly assuming his role in the family, Ross feels replaced as the father to his son. Struggling to play a role in his son's life, Ross must deal with these apparent failures as both father and husband. Seemingly unable to satisfy his wife or control his children, Ross, like Martin, maintains these little qualities throughout.

In this manner, Martin Tupper and Ross Geller are both quite Mitty-like. Experiencing the same powerlessness that characterizes this other little man, Martin and Ross both struggle in a world not unlike Walter Mitty's hell. Whether in the shadow of godlike men or lesbian lovers, the masculinity of Martin and Ross is constantly called into question. Likewise, like Walter Mitty, Martin and Ross maintain powerful dreams for their lives but somehow always manage to fall short. From settling for pornographic film success to living in the shadow of one's wife and her lover, these dreams constantly backfire, highlighting the littleness of these little men. Like Mitty, Martin and Ross seem limp, humiliated, and defeated.

While maintaining these very Walter Mitty-like characteristics, however, both Martin and Ross illustrate qualities in stark contrast to Mitty's persona. In an early review, for instance, Martin's *Dream On* was described as a "sexy adult comedy" and a "hip, risqué show" (Coe, 1990, p. 75), likely referring to Martin's ongoing sexual antics. His powerfully highlighted sexuality, indeed, seems contrary to the little man character of Walter Mitty. Martin sleeps with young coeds half his age (in an early episode he has a heart attack while sleeping with a 20-year-old college student), supermodels, rock stars, the most stereotypically beautiful women on television. In one episode, Martin actually breaks up with his rock-star girlfriend because he fears her heroin habit will be a bad influence on his son. That Martin can himself leave such a beautiful, famous woman heartbroken seems to counter the helpless, asexual characteristics of Walter Mitty. Not only does Martin sleep with stereotypically attractive women, but he also sleeps with different women almost every episode. In this, he is both sexually successful and adventurous as well. His adventuresome sex antics include sleeping with his son's teachers, asking out mothers of young women he has dated, dating women for whom he works, dating authors of books he is editing, and dating the young pastry seller in his office building. Episode after episode, Martin Tupper continually pursues these playboy antics. Far from Walter Mitty in his sex life, Martin appears to be a hypermasculine stud whose sex life is anything but little.

In the same way, Ross Geller also illustrates a sexy side that runs counter to Mitty's asexuality and littleness. While Ross may not sleep with a different woman every episode, as does Martin, his sexiness is repeatedly evidenced in popular discussions of his character. "Ross Geller is angsty, vulnerable and as cute as a wet retriever pup," Gerri Hirshey (1996) maintains. "Much of female America—especially the coveted 18- to 49-year-old free-spending, urban or fashionably suburban demographic dream gals—is ready to step up with a fluffy towel, a smooch and a soothing double latte" (p. 234). Ross Geller, Hirshey maintains, is the perfect sensitive, sexy '90s man, a far cry from Walter Mitty. Curiously, perhaps, David Schwimmer is repeatedly conflated with his character Ross. About her walk with Schwimmer during their interview, for instance, Hirshey comments that his "Rossian cluelessness sets us wandering amid dim sum parlors and Milanese bondage-wear shops" (p. 235). Kahn (1996) comments similarly, "Female fans equate [Schwimmer] with his *Friends* character, Ross Geller, a lovable paleontologist who has a history of being luckless in the romance department" (p. 50). Marin (1995) argues that "this symbiosis of actor and character may be what's made Schwimmer, 28, the breakout star of the show" (p. 68). In any case, asserting this symbiosis, popular discourses continually discuss Schwimmer and Ross' masculinity as one and the same.

Sheryl Kahn (1996), for instance, comments of Schwimmer:

> Exactly what has taken him from sitcom star to sex symbol in less than a year? With his boyish features and stubborn cowlicks, he's hardly hunk material…. Yet there's no denying his appeal. Playgirl voted him one of 1995's 10 sexiest men, and an on-line fan club spends hours waxing poetic over his puppy dog eyes. "Brad Pitt, David Schwimmer. Go figure," the 29 year-old laughs. "I guess I'll never understand it—I'm so boy next door." Which is precisely his charm—Schwimmer seems to be the kind of guy any woman would happily bring home to Mother. (p. 48)

In short, Ross (Schwimmer) is depicted as the sexy '90s man. "Tall, tentative Ross is the most conflicted. He's the melancholy X-er, Hamlet-like in his agonizing over whether it's manly for a guy to use fabric softener" (Marin, 1995, p. 68). Torn between his angst-ridden agonizing side and his sexy playboy side, "to many female fans the vulnerable, hangdog guy on *Friends* is the ideal mid-'90s man" (Marin, 1995, p. 68). In contrast to Walter Mitty's character, Ross, like Martin, is a sexualized, sexy man for the '90s. Far from the fully controlled, dominated "little man" of Mitty, Ross, like Martin, "may look geeky, but the guy's no wimp" (Marin, 1995, p. 68).

Martin Tupper and Ross Geller thus exemplify a sexual quality in stark contrast to Walter Mitty. Whereas Mitty seems pitiful and powerless throughout, Martin and Ross are presented as sexually potent, despite the powerlessness that characterizes the rest of their lives. Although Martin seems impotent in the shadow of Richard Stone, this does not keep him from sleeping with a host of the most beautiful women. Similarly, while Ross' failed relationships recall Mitty's failed dreams, he is still highly sexualized, viewed as the sexy mid-'90s man. In light of contemporary discussions of the crisis of masculinity, this hypersexualization of the little man proves an interesting move to negotiate conflicting notions of masculinity. Bringing this new sexuality to the character, these two sitcom personas create a little man for the '90s. While they may, indeed, dream lives they cannot completely fulfill, suffer from angst associated with these failed dreams, and evidence a certain lack of control over their everyday lives, Martin and Ross nonetheless come off as sensitive, sexy men, a far cry from Mitty's pitiful, completely helpless character.

Hypersexuality as Homophobia

It is telling that the issue of *GQ* that discusses David Schwimmer and Ross also contains an article entitled "'You Just Know He's Gay': Rumors about Celebrity Sex Partners Have Taken a Twist: We've Become a Nation Obsessed with Finding out Who's Bedding Whom ... of the Same Sex." In "a nation obsessed with finding out who's bedding whom ... of the same sex," this hypersexualization of the contemporary little man may indeed be a way to protect him from this obsession. In other words, Martin...constant sleeping around and the constant discussions of Ross' sexiness (that "he's no wimp") offer ways of highlighting these characters' heterosexuality. "As sensitive as they are," these characteristics seem to indicate, "this is a very heterosexual sensitivity."

This protection of heterosexuality indeed seems in line with much current discussion of the contemporary crisis of masculinity discussed above. For instance, while the traditional American male may have been conceived of as Goffman's "young, married, white, urban, northern, heterosexual Protestant father of college education," this contemporary crisis seems to call these characteristics into question. From growing movements in support of women's rights to increasing tolerance for alternative lifestyles,[11] a variety of discourses seem to counter the hold of this traditional masculinity. Indeed, Ian Miles (1989) maintains that "sexuality is an important issue in several of these movements, especially in those championing gay liberation and homosexual rights," stressing that these "are clearly difficult to accord with traditional masculinity" (p. 52). Given these challenges to masculinity, the hypersexualization of the little man offers a manner of resituating these character amidst this contemporary crisis, especially amidst issues of contemporary sexuality.

Supporting this idea, Judith Butler (1990) reiterates that "not to have social recognition as an effective heterosexual is to lose one possible social identity and perhaps to gain one that is radically less sanctioned" (p. 77). The characters of Martin and Ross both reposition the little man to deny the possibility of this less sanctioned social identity. In Martin's highlighted desire for women and Ross' highlighted desirability to women, both of these characters are deeply embedded in the heterosexual matrix addressed by Butler. "Indeed, the woman-as-object must be the sign that he not only never felt homosexual desire, but never felt the grief over its loss. Indeed, the woman-as-sign must effectively displace and conceal the preheterosexual history in favor of one that consecrates a seamless heterosexuality" (pp. 71–72). In the midst of discourses of masculine crisis, the consecration of this seamless heterosexuality appears all the more important.

While the little man may be seen as a desirable, even sexy, character for the '90s, these popular characters evidence an uneasiness not unlike Robert Bly's anxiety towards the soft male. Hypersexualized in order to highlight his heterosexuality, the '90s little man seems troubled by his own sensitivity. In this way, the hypersensitivity that rings folksy for Walter Mitty rings homophobic for Martin and Ross. Uncomfortable within contemporary discourses of masculinity, Martin and Ross prove negotiations of these competing, conflicting voices. While they champion the sensitive man of the '90s, they do so somewhat cautiously, situated firmly within this contemporary crisis of masculinity. They might be dreamy and hypersensitive, but they're no wimps.

These same tensions are evident, to one extent or another, in other 1990s situation comedies as well, themselves illustrating this same masculine crisis and heterosexist anxiety. On *Seinfeld*, for instance, characters Jerry and George both have to defend themselves from challenges to their sexuality. "Outed" by a magazine reporter, these characters spend an entire episode trying to prove their hyper-heterosexuality. Hoping to affirm their social identities in the midst of these conflicting discourses, Jerry and George engage in overtly heterosexual behaviors, repeatedly denying that they are in the least bit homosexual. "Not that there's anything wrong with that," they quickly add. Like Martin and Ross, these characters also evidence a contemporary masculine anxiety, demonstrating the widespread articulation of this contemporary crisis.

Indeed, little men such as Walter Mitty, Martin Tupper, and Ross Geller illustrate a sort of distressed masculinity, though one distinctly different in its contemporary version. Mitty feels himself disempowered, and attempts to gain power by dreaming of the masculine hero he might become. Similarly, Martin and Ross are also disempowered, controlled by the circumstances of their lives, their ex-wives, and their ex-wives' lovers. However, Ross and Martin's hypersexuality illustrates a relationship to the contemporary crisis of masculinity that is profoundly different from Mitty's character. Enid Vernon (1976) argues that "in a profound sense, the little soul is the American Superman switching identities, as it were, in an imaginary telephone booth" (p. 203), and this '90s little man seems to maintain a superidentity of his own, engaging the character developed with Walter Mitty in a distinctly '90s way. Both little and hypersexual, these dynamic characters anxiously weave their way through the contemporary crisis of masculinity.

Little Big Man Clinton

From the Clinton presidential campaign to action heroes and little men, the selection of '90s men above illustrates an attempt to address contemporary critiques of masculinity by offering a new, more sensitive male. In each case, these contemporary deployments of masculinity work against and alongside preceding representations, their producers seeking to distance them from an old masculinity even as they implicitly espouse it. Clinton, for instance, both reacted to and reworked the imagery of the Reagan/Bush presidency, offering a new (male) President that stressed sensitivity and empathy, rather than toughness or strength. Here, Clinton challenged a set of previously established notions of presidential masculinity, but in a way generally seen as positive (judging by election results) by the American public. By offering America a newly sensitive, empathetic presidency, the

Clinton campaign promised an administration distanced from the oppressive masculinity associated with the '80s.

At the same time that Clinton defiled the preceding version of presidential masculinity, the Clinton campaign also paid homage to the same Reaganite vision of the masculine president, working, like the producers illustrated above, to hold onto a traditional masculinity even as they appeared to abandon it. Just as Steven Seagal would eventually kick the ass of an offending oil worker—offsetting his new male sensitivity—so Clinton would continually attempt to illustrate his own ass-kicking potential. Clinton's campaign video for the 1992 Democratic National Convention, for instance, had Clinton, as well as his mother, telling the story of young William's firm stance against his abusive stepfather. Even as an adolescent, the biographical video seemed to say, Clinton could hold his own, alleviating any fear that his empathy and sensitivity might be translated as weakness. Similarly, Clinton was hypersexualized in a way not unlike little men Ross and Martin, ironically, however, not by his own "producers," but by his archenemy Kenneth Starr. Indeed, Clinton shared many characteristics with Ross and Martin. "Struggling" as the underdog in the '92 election (discussed in more detail in chapter 1), husband to the outspoken Hillary (who was stereotyped as a lesbian in much popular discourse), and living a presidential life in the shadow of great man JFK, Clinton is positioned with these other dreaming little souls of the '90s. As with Ross and Martin, Clinton's hypersexuality clearly displaced any possibility that his is anything but a purely heterosexual sensitivity. Like these other "sensitive '90s men" Clinton both defiled a conventionalized masculinity and worked against this defilement.

It is important, of course, that these perceptions of Clinton were not all within the hands of himself and his producers. Whether via the Starr Report, *Time*, or a *Tonight Show* monologue, the discourses surrounding President Clinton did as much to establish his conflicted sensitivity as his own administrative rhetoric. These multiple discourses evidence still further the popular tensions Clinton simultaneously negotiated, fell victim to, and perpetuated. As both a source of this sense of conflicted sensitivity, via his administration's attempts to position him as both sensitive and tough, and a site where others articulated these same tensions, Clinton's public persona demonstrated the dynamic nature of this '90s' struggle over masculine sensitivity. Just as Clinton defiled an older presidential masculinity, only to paradoxically celebrate many of its same qualities, so Kenneth Starr used these same tensions and anxieties in his attempts to discredit Clinton.

In the end, each of the men discussed above worked to negotiate the conflicted masculine sensitivity of the '90s, offering a new, sensitive '90s male, only to reiterate and reinvigorate much of the anxieties they purported to abandon. Just as Clinton invoked a tension between empathy, sensitivity, and toughness, only to see that same tension come back in the form of the Starr Report, so each of the masculine representations above both hold onto and rebuke a new '90s sensitivity. Ultimately, even as they seem to defile a previously established conception of masculinity—the president, the action hero, the leading man—these new deployments of masculinity worked to reproduce conventionalized masculine values and anxieties, but in the subtly different form of the new, sensitive male.

Notes

1. See Kellner (1995), chapter 2, for a like discussion of Rambo-era masculinity and its connection to Reaganism.

2. Rotundo is here speaking of masculine stereotypes of the late 19th century. However, such idealizations of violence and fighting are equally apparent in contemporary notions of masculinity as well. Jansen (1994), for instance, discusses values of hegemonic masculinity "(i.e., aggression, competition, dominance, territoriality, instrumental violence)" (p. 10), evident in "sport/war metaphors" of the Persian Gulf War.

3. I here use "Seagal" as synonymous for any character within his films, for, surely, it is the personality and not the single character that drives the Seagal persona.

4. For a further discussion of this film, see Ross (1995).

5. See Russell (1994).

6. See Russell (1994).

7. Garber's book, *Vested Interests*, offers a detailed discussion of African American masculinity and transvestitism in its chapter 11 (pp. 267–303). Likewise, see Russell (1994).

8. I do not mean to equate Bly's soft male with the little man of Mitty, Martin, and Ross, although these notions may indeed have similarities. I merely hope to emphasize the ways in which both ideas demonstrate a notion of masculine crisis.

9. For further discussion of Mitty's character and of Thurber's little man, see Kaufman (1994), Cheatham (1990), and Blythe and Sweet (1986). Likewise, see Gopnick (1994) for a discussion of Thurber's use of comedy as "deflation," and Vernon (1976) for a discussion of the development of the little man in American comedy and satire.

10. Kaufman, though, himself argues more strongly that "Mitty's rejection and withdrawal from the world are more radical than can be denoted by the idea of 'daydreaming.' In fact, we witness the descent of Mitty into ever increasing preoccupation with his fantasy life and increasing rejection of the so-called real world. His withdrawal is symptomatic not of mild-mannered exasperation with a trivial world, but of anger and misanthropy" (p. 93).

11. For a discussion of public discourse around gay liberation, for instance, see Darsey (1994, 1981a, 1981b).

Chapter 3

Classified and Declassified:

Cultural Capital and Class Anxiety in New Male Sons and Fathers

President Clinton's sensitivity was not the only component of his conflicted character—just as conflicted sensitivity was not the only tension to be negotiated among these '90s new men. Born in Hope, Arkansas, and thus associated with the rural South, and yet educated at Georgetown, Oxford, and Yale University Law School, Clinton's character maintained a conflicted class status as well, a tension evoked by the mainstream media, Clinton's detractors, and the Clinton campaign itself. Discussing one of Clinton's campaign appearances, an article in *Business Week* quipped, "Clinton, who had been buffed to a high sheen at Yale Law School and Oxford University, stepped before a throng of Miami supporters and described himself as 'an ol' redneck.' Aw, shucks, it worked" (Walczak et al., 1992, p. 26). At the same time, a *Washington Post* article called him a "would be President of both pie charts and moon pies," commenting that "Clinton seems to have more facets than a debutante's engagement ring," that is, "a little bit Tiffany diamond" and "a little bit Woolworth rhinestone" (Von Drehle, 1992, p. 1).

In this juxtaposition of educated Oxford scholar and down-home country redneck, Clinton's persona participated in an interesting class dynamic that is important to traditional masculinity as well as to the supposed new man of the '90s. On the one hand, as suggested by Goffman's unblushing male in America, for the better part of the 20th century a certain level of middle-class success has been one of the dominant ideals of American manhood. Ward Cleaver and Ozzie Nelson model a middle-class rationality and suburban comfort indicative of this unblushing middle-class man (a lifestyle of conspicuous consumption noticed earlier in the century by Thorstein Veblen). Well employed and endowed with middle-class sensibilities and taste, Ward and Ozzie appreciate the manliness of imported cigars, well-decorated dens, and dry-cleaned suits. Even the '50s "Playboy," Hugh Hefner's rebellious answer to these '50s family men, reveled in the possibilities of middle-class or better consumption. Indeed, Hefner himself celebrated the Playboy's sophisticated appreciation of "mixing up cocktails and an hors d'oeuvre or two, putting a little mood music on the phonograph, and inviting a female acquaintance for a quiet discussion on Picasso, Nietzsche, jazz, sex" (Hefner, December 1953, p. 3).[1] In contrast, the unsophisticated working-class man of the '50s, such as the *Honeymooners'* Ralph Kramden, remained hilariously unrefined, his ever-present bus driver's uniform little match for the power suits of Ward and Ozzie or Hef's slippers and smoking jacket.

Even as 20th-century manliness has relied on a middle-class sensibility, however, this same middle-class sophistication has likewise been seen as antithetical to the rugged individualism that also dominates conceptions of American masculinity. Images of the 1950s "man in the gray flannel suit," for instance, who spent his days pushing papers in an office, demonstrate a sense of uneasiness towards the supposed sophistication and comfort of middle-class manhood. In contrast to the celebratory images of Ward and Ozzie, this white-collar worker was also imagined as a corporate drone whose identity had been lost in the nine-to-five workday of the middle-class world. Wrapped up in the comfortable world of offices and manicured lawns, the gray flannel suits were seen as effeminate yes-men whose masculinity had been compromised by corporate and suburban conformity. Frank Stark, Jim Backus' character in *Rebel Without a Cause*—the father to James Dean's rebellious Jim Stark—illustrates these anxieties clearly. An early scene has Frank dressed in an apron and cleaning up a tray of spilled food. "What can you do when you have to be a man?" Jim asks, interrupting Frank's nervous, organization man's ramblings: "I'll get some paper and we'll make a list. Then if we're still stuck, we'll, we'll get some advice."

Given this simultaneous celebration and rejection of middle-class sophistication, dominant notions of masculinity have had an uneasy relationship with working- and lower-class masculinities. On the one hand, working- and lower-class masculinities are mocked for their presumed violations of middle-class masculine sensibilities. Stereotypical working-class men Ralph Kramden and Archie Bunker (*All in the Family*) are excessive in both body and speech. Overweight and subject to consistent, uncontrolled verbal outbursts, their inability to control themselves guarantees their place as butt of each episode's joke. Alongside the middle-class, unblushing male, they are hilariously uncivilized and uncouth. At the same time, however, popular culture regularly celebrates this same violation of middle-class civility as an anecdote for middle-class comfort and oversophistication. In advertisements for Ford, Dodge, and Chevrolet, a "real American masculinity" repeatedly draws on images of men who know how to get their hands dirty and would rather experience the world than push papers across it.

President Clinton's conflicted character illustrated these same schizophrenic negotiations of masculinity and class. On the one hand, Clinton's Hope, Arkansas, roots, stressed by the Clinton administration throughout its 1992 campaign, allowed for Clinton's association with lower-class experience, distancing him from the cold world of real (that is, bureaucratic, heartless, moneyed) politics. A July 1992 article in *Maclean's* depicted Clinton as "forged by a southwestern upbringing far from the now tainted world of Washington politics," quite different from Bush: "a president whose roots are imbedded in upper-class America—the scion of a wealthy Connecticut family who came of age in the Houston business establishment and who is now the consummate Washington insider" (Wallace, 1992, p. 27). At the same time, Clinton's own excessiveness was also a source of conversation throughout his presidency, demonstrating the sort of "convivial indulgence"[2] typically associated with—and mocked in—lower-class men. Describing Clinton's excessive love of fatty food, a 1992 *New York Times* article notes, "From Sims Bar-B-Q to Junanita's, from Doe's Eat Place to Hungry's Café, President-elect Clinton prefers the stuff with fat in it: jalapeno cheeseburgers, chicken enchiladas, barbecue, cinnamon rolls and pies" (Burros, 1992, p. 1). Similarly, "Governor Jack Stanton," John Travolta's character in the film *Primary Colors*, and a hardly subtle reference to Bill Clinton, spins such wonderful lines as "rube-ass, bare-foot dip-shit third-rate Southern fried piece-of-shit Alderman," and eats more powdered donuts than would seem physically possible. In such moments, Clinton's "working-class experience" demonstrated a conflicted character like the one discussed above, Clinton depicted variously as a man of the people and an excessive consumer of powdered pastries.

This chapter works to understand how these class tensions are negotiated within the contemporary '90s man of the mainstream media. Given the already problematic relationship between dominant masculinity and working- and lower-class masculinities, how are these class tensions articulated during a period of masculine crisis—at a moment in which representations of masculinity have become particularly self-reflexive? This chapter focuses on fathers and sons, understanding how these particular presentations are articulated within a climate of '90s masculinity. The delusional imagination of Reagan's '80s offered up a series of young men with expendable cash and middle- to upper-middle-class sensibilities—the well-off sons of Reagan. From *Ferris Bueller's Day Off*, in which young Ferris skips school to spend a day driving around Chicago in his friend's dad's Ferrari, to *Weird Science* and *Revenge of the Nerds*, in which young nerds prove their self-worth through their brains and computer technology, these Reagan sons used their money to win popularity, approval, and, of course, women. Likewise, although Reagan dads Stephen Keaton (*Family Ties*) and Michael Steadman (*thirtysomething*) worked jobs that couldn't always guarantee a steady income—public access television and freelance advertising, respectively—they nonetheless maintained their status as loving, nurturing middle-class dads throughout, proof, it seemed, of the fidelity of Reaganomics and trickle-down family values.

Relative to these images of Reagan's '80s, the '90s fathers and sons discussed in this chapter demonstrate a reintroduction of working- and lower-class men, from the parodic images of Homer Simpson and Beavis and Butthead to Jack Dawson, Leonardo DiCaprio's character in *Titanic*. In concert with their contemporary ideas of masculine crisis, these '90s men negotiate middle- and lower-class masculine sensibilities in conflicted and bizarre manners, distancing themselves from the middle-class men of Reagan's '80s like the sensitive male from the Reagan hard body. Discussing a series of "working-class films" of the '70s—another moment in which lower-class men served as mediators for anxieties within middle-class masculinity—Peter Biskind and Barbara Ehrenreich (1987) argue:

> In the films of the late seventies, the (previously invisible) working class be-
> comes a screen on which to project "old fashioned" male virtues that are no
> longer socially acceptable or professionally useful within the middle class—
> physical courage and endurance, stubborn determination, deep loyalty
> among men. (p. 206)

If '70s films such as *Bloodbrothers*, *Saturday Night Fever*, and *Rocky* (all
discussed by Biskind and Ehrenreich) hold up working-class men as a
screen for "old-fashioned" masculinity, the '90s images discussed in
this chapter hold up a broken lens—reflecting and refracting their
working-class images through the contemporary crisis of masculinity.
Whether through hyperbolic parody, as in the case of Homer Simpson, or through a presumed "celebration" of "working-class experience," as with Jack Dawson, these deployments of masculinity endow
their working-class images with a sense of middle-class cultural capital, holding working-class manhood at arm's length even as they
celebrate its "new male potentials."

New Male Idiots: Parody as Middle-Class Distance

Mark Crispin Miller's essay, "Deride and Conquer," written in 1987,
offers an interesting history of the situation comedy, placing the '80s
sitcom and sitcom family within a larger history of advertising and
consumption. For Miller, the continued proliferation of advertising
and the rapid-fire semiotics of consumption that characterize
Reagan's '80s have led to the commercialization of American fantasy.
With the proliferation of music videos, infomercials, and product
placement blurring the line between entertainment and advertisement, television has become "pervasively ironic," flattering its viewers with a sense of choice and freedom that is precisely the freedom to
consume:

> Before the eighties an ad for Pepsi, or for some comparable item, would
> have worked differently—by enticing its viewers toward a paradise radiat-
> ing from the product, thereby offering an illusory escape from the market
> and its unrelenting pressures. In such an ad, the Pepsi would (presumably)
> admit its drinker to some pastoral retreat ... [which] would retain its other-
> worldly charms.... Now products are presented as desirable not because
> they offer to release you from the daily grind, but because they'll pull you
> under, take you in. (p. 188)

Miller sees *The Cosby Show* as the penultimate celebration of '80s consumption, a half-hour commercial for artwork and colorful sweaters aplenty that invites us to suspend our disbelief and feast our eyes on the good life. According to Miller, "*The Cosby Show* attests to the power, not of Dr. Cosby/Huxtable, but of a consumer society that has produced such a tantalizing vision of reality" (p. 210). In a world inundated with celebrations of consumption, consuming itself has become a sort of escape, the very form of the American fantasy of Reagan's '80s.

Miller's discussion of the pervasive irony of the '80s prophesies a hyperbolic parody that pervades the '90s sitcom and its presentations of masculinity. If advertisements and sitcoms of the '80s predominantly invited viewers to suspend their disbelief, asking them "not just to buy its products, but to emulate them, so as to vanish in them" (p. 224), a series of '90s ads seemed to ask viewers to reject products, yet consume them nonetheless. In contrast to the Pepsi ad that Miller discusses, which invites the viewer into a fantasy world that is not otherworldly, but Pepsi-worldly, a late '90s Sprite campaign claims "image is nothing, thirst is everything," seemingly shattering the entire fantasy of consumer escape with a knowing recognition: "Hey, it's just a drink." An MTV advertisement ironically urges viewers to "Free Your Couch," as if the network would prefer viewers spend their time outdoors, casting MTV, ironically, as the network for people who don't watch television.

With a similar hyperbole, while the sitcoms of the '80s may have served as ads for the good life, celebrating Reagan's '80s with living rooms overcrowded with consumables, shows such as *Beavis and Butthead, Married ...with Children*, and *The Simpsons* offer parodic versions of these shows, their sparse, dingy houses, dysfunctional relationships, and mocking language jeering backward, it seems, at Reagan's '80s. These next sections work to understand the changing conceptions of masculinity that accompany these parodic programs of the '90s. While the oafish working-class male, or troubled, delinquent youth, are not original to the sitcoms of the '90s, the popularity of new sons Beavis and Butthead, and new dads Homer Simpson and Al Bundy, make them particularly interesting sites for exploring '90s masculinity. Fathers and sons of MTV (*Beavis and Butthead*) and Fox (*Married ...with Children*, and *The Simpsons*), these men are indicative of a movement to appeal to a generation of viewers assumed to be particularly savvy about media images and messages.[3] Here, the irony that Miller observes in the '80s sitcom reaches hyperbolic proportions, creating peculiar attempts to address these media-savvy viewers. Celebrated for their stingingly parodic critiques of the '80s family, these programs ultimately offer an uncomfortable negotiation of middle- and working-class masculinities, their celebrated critical humor proving a hyperbolic irony that holds working-classness at a safe, middle-class distance.

Adult Adolescents: Beavis and Butthead

Since their arrival on the public scene, cartoon characters Beavis and Butthead remained the subject of much criticism and public debate. With incidents of cat burning and other acts of juvenile mischief associated with the show, its two fire-loving delinquents became easy targets for critics of the state of the day's youth. Yet, more interesting than these criticisms are the widespread celebrations of the show's parodic potentials, its celebrated ability to savagely mimic the disaffected young male of the '90s. When Beavis and Butthead aren't destroying animals or property, they spend their time watching music videos and offering their own wise commentary on their contents, typically "This sucks!" or "This is cool!" crafting a picture of the music video generation that many have celebrated for its satiric qualities. According to one celebrant:

> "BandB" is so crude in its satire, so unrelenting in its implicit critique of the couch-potato mind-set, so severe in its condemnation of everything Beavis and Butthead represent, that I—in all honesty—must regard it as a moral act. If I had a kid of twelve or so, I'd much rather he watched "BandB" than, say, the softcore and softhead blandishments of "Baywatch" or "Beverly Hills 90210." (McConnell, 1994, p. 30)

Further stressing the apparently savage critique leveled by *Beavis and Butthead*, an article in *National Review* argues that "One thing is certain: you simply cannot like Beavis and Butthead and also like most of what MTV represents" (Gardner, 1994, p. 61). Still another writer claims, "this may be the bravest show ever run on national television: it lampoons not just the performers who are the channel's raison d'être, but mercilessly depicts MTV's dopey, anti-social, suburban-boy audience" (Andersen, 1993, p. 75). Beavis and Butthead seem the most ironic critics of '90s culture, all the more so because they critique the very network that gives them life.

The show itself, of course, does its best to highlight and exploit this sense of parody. Beavis and Butthead are pure id. Working to satisfy one insatiable desire after another, they both embody the most stereotypical adolescent male passions—food, beer, sex, rock-and-roll, and, of course, television. Dressed in their concert shirts and seated in front of the television, Beavis and Butthead are the embodiment of the quintessentially disaffected youth. Butthead seems the more intellectual of the two characters, which is to say he has the most lines. His constant misinterpretation and mishearing of names and phrases illustrate an adolescent mind obsessed with sex, a hardly subtle commentary on the pubescent male. When a teacher instructs, "There's an exciting world when we discover that we don't need TV to entertain us," Butthead laughs. "He said anus." Likewise, the principal's name, McVicker, translates easily to Principal McDick-her. In the feature film, appropriately titled *Beavis and Butthead Do America*, Beavis and Butthead's obsession with "doing it" leads them on a trip across the country, taking them as far west as Las Vegas (where Beavis hopes to meet lots of "sluts," his mishearing of "slots") and as far east as Washington, D.C. On the plane to Las Vegas, Beavis devours the candies, breath mints, and caffeine pills from the purse of an elderly woman, sending him on a sugar high that nearly wrecks the plane. Desiring, hormonal teens, Beavis and Butthead embody a stereotypical image of the adolescent male—the basis for the biting satire that critics recognize and applaud.

One particularly interesting episode of the television program, titled "Generation in Crisis," has a graduate student visit Beavis and Butthead's high school classroom in order to make a documentary film about disaffected youth. As the episode opens, Beavis and Butthead sit in the back of the classroom exchanging blows to the head, Beavis hitting Butthead with a ruler and Butthead hitting Beavis with a textbook. Interrupting this mayhem, their teacher introduces Ken Alder, a graduate student in film and anthropology and a practitioner of documentary filmmaking, which Alder describes as "the fine art of capturing real life on film." When Alder announces that the film, *Generation in Crisis*, will be "about so-called trouble kids," the class quickly and unanimously identifies Beavis and Butthead. The next scenes have Beavis and Butthead walking down the street with Alder framing them for the camera, waiting for them to become the disaffected youth we all know they are. As oblivious to the camera as they are to the rest of the world, however, Beavis and Butthead stare blankly ahead, as Alder begs them in a frustrated tone: "Engage in some antisocial behavior or something."

The resulting "film" makes clear the hyperbolic parody in which *Beavis and Butthead* takes part. Presented as a "black-and-white film," with the "film screen" given as part of the picture, this cartoon film immediately parodies the documentary genre it pretends to reproduce. The opening scene "shot" through the window of a car as it drives down the street, Alder begins his voice-over: "It could be any American town. Hardworking men and women go about their daily lives—raising families, pursuing the American dream." The voice-over pauses as the car moves along the street, taking in this typical American town. "Where are the children?" Alder asks, continuing his voice over. "You're a wuss!" Beavis' voice interrupts. "No way butt munch, you're a wuss!" Butthead replies. "No way, you're a wuss," Beavis continues. Continuing his voice-over, Alder explains:

> We'll call them Steven and Bernard, although those are not, in fact, their real names. In the wealthiest and most scientifically advanced nation the world has even seen, they lead an existence devoid of meaning and barren of intellectual content.

With the words "barren of intellectual content," the scene cuts to a picture of Beavis and Butthead eyeing a dead armadillo on the street. "It's flat," Butthead astutely observes. "Huh, Yeah. It's flat and it's dead," Beavis contributes. Noting the obvious, Alder adds, "Their primary influences seem to be a steady diet of bland television and loud, mindless, heavy metal music."

Ever hyperbolic in its parody, here, *Beavis and Butthead* simultaneously mocks the adolescent mentality that likes roadkill as well as a graduate student's artsy documentary film that might find this worthy of reproducing or reflecting upon. "Tempting though it may be," continues Alder, "it is not for the documentary filmmaker to pass judgment on his subjects. Let us hear from Steven and Bernard in their own words." Highlighting his comical academic-speak and overintellectual attempt to dig below the surface, Alder asks Beavis, "In a word, Steven, what is your raison d'être?" making use of his best overly affected French accent. "Uh, it's in my pants," Beavis re-

sponds, demonstrating his own phallic fixation and misunderstanding of the question. "Where do you picture yourself in ten years?"
Alder asks Butthead, who stands behind a counter dressed in his
Burger World uniform, the "W" on his hat a thinly veiled reference to
McDonald's golden arches. When Butthead doesn't respond, Alder
lectures, "You know the 21st-century marketplace you're going to enter will be a global electronic village. How are you preparing yourself
for what's bound to be a complex and challenging world?"

While this wordy diatribe again mocks Alder's highbrowed intel-
lectualism, Butthead's laughing response, "Uh, you said enter," pokes
fun of this adolescent stupidity as well, continuing this dual, hyperbolic parody.

The hyperbolic irony of this cartoon mock-documentary affords a
doubly distanced stance, a perfect example of the critical distance
celebrated by the writers quoted earlier. When Alder asks Beavis,
"What are you feeling right now?" Beavis responds, "My left nad. It
itches. Can I say nad?" To which Alder, increasing the ironic distance,
replies, "There's no censorship of any kind. This is an independent
documentary film. With a generous grant from our friends at the Exco
Corporation." This final joke—the independent film that is not one—
relies upon a particular kind of cultural capital that is familiar with
these "artistic films." Making fun of both the hyperintellectualism of
Alder and the hyperstupidity of Beavis and Butthead, *Beavis and Butthead* offers a middle ground of middle-class common sense and distance, laughing at and with Beavis and Butthead at the same time that
it laughs at and with Alder. Maintaining a hyperbolic distance that
gets the play on McVicker's name at the same time that it understands jokes about documentary filmmaking, *Beavis and Butthead*
maintains a comfy middle-class cultural capital that gets the jokes
without succumbing to them.

But the joke, ultimately, is on the adolescent welfare boys that
Beavis and Butthead satirically reproduces, the unspoken references in
this parody of boys' life. Beavis and Butthead are both fatherless and
their mothers only appear in the boys' sexual references and jokes on
each other. Their "fathers," we learn in *Beavis and Butthead Do America*, were roadies for a band that passed through town, themselves
irresponsible adult adolescents who care only about "scoring with
chicks" ("That's cool!" Beavis and Butthead respond in unison). Sons
of irresponsible parents, Beavis and Butthead embody a '90s fear of
the lower-class adolescent male. Without the proper parental author-

ity, the story goes, the adolescent male "grows up" before he's ready. Beavis and Butthead's enlarged heads, too big for their boys' bodies, seem symbolic of this premature growth. Unable to handle his developing impulses, the overgrown adolescent male turns his destructive hormones towards the community. But whereas Dylan Klebold and Eric Harris of Columbine infamy were from middle-class, nuclear families, Beavis and Butthead reside permanently in the lower, welfare classes — the students their classmates quickly recognize as "so-called problem kids." Here, the threat of the "adult adolescent male" is contained, held at a parodic distance from which Middle America can jeer at his hyperbolic mischief.

In this way, even amid these parodic critiques in which *Beavis and Butthead* lashes out at its symbolic father, MTV, poking fun at the network, its program, its viewers, and the contemporary world, Goffman's male still manages to rear his unblushing head. While Theo Huxtable and Alex P. Keaton, of '80s sitcoms *The Cosby Show* and *Family Ties*, celebrated middle-classness with their dress, mannerisms, speech, and being, *Beavis and Butthead* do so through a hyperbolic irony that holds up the welfare boy to derision, laughing with, alongside, and at his lower-class cultural capital. In the process of all of this, *Beavis and Butthead* reinforces a version of middle-class common sense of importance to the unblushing male in America.

Adolescent Adults: Al Bundy and Homer Simpson

In a similar way, sitcom characters Homer Simpson and Al Bundy feign an interesting '90s parody. Hyperbolic in ways similar to Beavis and Butthead, Al and Homer have themselves been celebrated as parodic reactions to the fathers of the '80s. While *Married ...with Children*, for instance, began in 1987, it is nonetheless a uniquely '90s program, its celebrated critique of '80s television lasting well into the '90s. A 1989 article in *Rolling Stone* explains the show's parodic force through the goals and words of its creators, Michael Moye and Ron Leavitt:

> Leavitt and Moye say their original inspiration for the show sprang from their hatred of sitcoms like *Family Ties*, which all too often devolved into laughless, weepy kitchen therapy sessions in which Dad and Alex finally mustered up the courage to trade *I love yous* because the family pet or

grandmother had died. "Is every family happy?" Leavitt asks. "Do they all dress well? Do both Mom and Dad have great jobs in every house in the world?" (Simms, 1989, p. 30)

A 1996 article in *Newsweek* continues this celebration, calling *Married ... with Children* "crass, low class and the longest-running show on TV":

> Think back to 1987. "The Cosby Show" was No. 1, the stultifying "Growing Pains" was in the top 10. "Family Ties" was the closest these saccharine up-scale fantasies came to edgy. Then "Married ..." came along, a breath of foul air in a roomful of Pine Sol.... They made nuclear waste of the nuclear fam-ily.... Instead of a father who knows best they came up with one who lies and smells bad. (Marin, 1996, p. 70)

In contrast to '80s dads Cliff Huxtable or Steven Keaton, Al is here celebrated as the smelly fall-out of the "dysfunctional family" talk of the late '80s and '90s, a satiric stab at the delusional marital and famil-ial bliss of Reagan's '80s. *The Simpsons* has been similarly celebrated in the mainstream media, seen as a like parody of the '80s good life. One writer observed early on that:

> "The Simpsons" is satire. Rather than engage in the pretentious misrepre-sentations of family life that one finds in the "model family" shows (from "The Donna Reed Show" to "The Cosby Show"), this program admits that most parents aren't perfect. They haven't worked out their own childhood confusion, and they don't have the answers to all their children's prob-lems.... *The Simpsons* show us in a rather bald as well as witty way what is was about our upbringing that made us brats as kids and neurotic as adults. (Rebeck, 1990, p. 622)

Both *Married ... with Children* and *The Simpsons* poke constant fun at the "functional families" of the '80s, the shows' fathers, Al and Homer, central caricatures in these satiric farces.

Of course, Al and Homer are not the first fathers to be satirized in this manner. In his history of the sitcom father, Mark Crispin Miller (1987) tells a tale of increasing derision tracing the decline of father from the Ward Cleaver images of the '50s to his contemporary, '80s dad. In the '50s "Dad seemed to reign supreme in sitcom country—or at least in its better neighborhoods" (p. 196). Here, "As the apparent ruler of the world, Dad, pushed around by no one, somehow deflated all his inferiors—and all were his inferiors" (p. 197). The dads of the '60s and '70s, however, begin to reveal cracks in this fatherly armor, the growing economic importance of Mom and the kids—who had become more important consumers—displacing Dad's paternalistic reign. In the '70s Archie Bunker, another lower-class dad of parodically extreme "–isms" (racism, sexism, nationalism, etc.), offered a figure with interesting parallels to Al and Homer. As Miller puts it:

> As usual, the lower-class father figure was presented as a joke, but now his subversion appeared politically correct. Archie Bunker was a butt, not because he was an overreaching loser, like Ralph Kramden, but because he was socially unenlightened—sexist, racist, militaristic, a nexus of reactionary attitudes that allowed his juniors and his viewers to deride him in good conscience. (p. 203)

Finally, the '80s dads, for whom Bill Cosby is Miller's spokesperson, have been reduced to a walking billboard of consumer products and the good life. Indeed, for Dr. Huxtable, Cosby's well-appointed Dad, "affluence is magically undisturbed by the pressures that ordinarily enable it" (p. 209). In the process, Miller continues, "Forever mugging and cavorting, throwing mock tantrums or beaming hugely to himself or doing funny little dances with his stomach pushed out, Cosby carries on a ceaseless parody of some euphoric eight-year-old." No different from the Cosby of Coke, Ford, or Jell-O, for which he advertises, the Cosby Dad is reduced to a playful child welcoming the audience into the fun world of consumer products.

Al Bundy and Homer Simpson carry this derision of daddy in interesting new directions, twisting the unblushing masculinity espoused by Cosby et al. The derision Miller notes with Cosby is the reduction from authoritative patriarch to beaming product pusher. Al and Homer illustrate a newly hyperbolic and excessive parody of dad. If Bill Cosby is an eight-year-old advertising consumer goods, then Al and Homer are adolescents whose appetites and habits of consumption are out of control, proof, it seems, of their excessive impulses of consumption. Homer Simpson's appetites for beer, candy, television, chili, bacon, and so on, are well established. In one episode, Homer steals a "Gummi Bear" in the shape of Venus de Milo ("carved by Gummi artisans who work exclusively in the medium of Gummi," the salesperson explains). "Must have rare Gummi," Homer drools, before he smashes the glass display case and makes off with it. At the annual chili cook-off, Homer drinks candle wax, protecting his mouth so he can eat several helpings of chili made from "the merciless peppers of Quetzlzacatenango! Grown deep in the jungle prime-val by the inmates of a Guatemalan insane asylum." After an arrest for drunk-driving Homer must attend Alcoholics Anonymous, con-fessing at one meeting: "The other day I was so desperate for a beer I snuck into the football stadium and ate the dirt under the bleachers." In each case his excessive appetite pushes him towards hyperbolic extremes of consumption.

Similarly, Al Bundy demonstrates his own share of ridiculously insatiable appetites. His constant references to his favorite magazine, *"Big Uns,"* and his regular trips to the "nudie bar" evidence an ado-lescent obsession with sex similar to that of Beavis and Butthead. On his son's eighteenth birthday he announces, "You're going to do what every male Bundy does when he reaches the age of eighteen.... To-night I'm taking you to the nudie bar!" While there, Al prances on stage with one of the dancers, helping teach son Bud a set of nudie bar rituals. "This place sucks when you're broke," Bud complains later, showing his boredom. "Every place sucks when you're broke," Al retorts. To which Bud asks, "So why do you come here?" In an-swer to Bud's question, Al picks a fight with a passing patron, pro-voking a room-clearing brawl meant to liven up their evening. Here, Al demonstrates his own hyperbolic appetites, his obsession with sex and fighting demonstrating a hyperconsumption similar to that of Homer. Introducing Bud to the ritual of the strip club by dancing on stage with strippers and picking fights with fellow patrons, Al seems to initiate Bud into this world of adult consumption—the elder intro-ducing the younger to this brand of adult adolescent behavior.

In this way, the hyperbolic consumption that both Homer and Al demonstrate twists the masculinity of the '80s on its head. The dad of the '80s was himself hyperconsumptive, serving up consumer goods aplenty though never seeming to struggle for this lifestyle. Homer and Al's consumption is all struggle, an excessive desiring after a host of stereotypically masculine goods, from sex to chili to beer. The idi-otic ways in which Homer and Al quest after their desires as well as the priority they accord to them offer a twisted parody of the pro-vider role played by the dad of the '80s. While Cosby and his fellow dads provided sweaters, food, "culture" (in the form of art work, jazz, and other highbrow entertainment), and love in abundance, Homer and Al provide almost nothing, and seem barely to care. The '80s dad knew where his priorities lay. Here, love and consumption were one and the same. To love one's family was to help them live the dream of

Reaganomics. For Al and Homer, however, these priorities are not so clear. Far from the '80s dutiful dad, they struggle to negotiate their priorities, feeling a powerful discord between their hyperconsumptive desires and the needs of their families.

Homer constantly disappoints the other Simpsons in the pursuit of his desires, the rest of his family seeming mere inconveniences in his otherwise happy life. Reading a computer file at the "Springfield Bureau of Records," Homer reprimands an employee: "Ah, hah! See!? This thing is all screwed up! Who the heck is Margaret Simpson?" To which the employee replies, "Uh, your youngest daughter." Homer's inability to remember his daughter Maggie, as well as his disregard for his children's safety—he regularly chokes son Bart for misbehaving—demonstrate this inattentiveness to family. In one episode, Homer's older daughter, Lisa, tries to get closer to him by taking an interest in football. When she asks to join him, Homer agrees, though with some obvious reluctance: "Just don't say anything and sit down over there." When Lisa sighs in disappointment to his "welcome," Homer protests, "Lisa, please, I can't hear the announcer." However, when Homer realizes that Lisa has a knack for picking the winning team, every Sunday becomes "Daddy-Daughter Day," with Lisa cast as Homer's ticket to big winnings with his bookie. At this point, Homer and Lisa enjoy the father-daughter bonding they had previously missed. When Lisa observes, "Houston's failed to cover their last ten outings on away turf the week after scoring more than three touchdowns in a conference game," Homer replies lovingly, "Oh, my little girl says the cutest things." Of course, when the football season draws to a close, Daddy-Daughter Day does as well. "Don't worry," he tells Lisa, "the new football season is only seven months away." Although Homer and Lisa eventually resolve their dispute, that Homer must struggle in this process demonstrates a lack of concern for family and a constant willingness to place his own desires over his love for the other Simpsons.

Al evidences a similar disdain for family that parodies the "good provider role" of the '80s. At the beginning of the episode discussed above, Al returns from work, lamenting his job as a shoe salesman and his life overall. "Well, I chalked up some more frequent loser miles today," Al begins. Discussing the specifics of his day, Al explains that the Department of Juvenile Corrections brought a group of juvenile delinquents into his shoe store, forcing them to watch him work as incentive to stay in school. "It's a new program called Scared Rich," Al explains. Quickly proceeding to a general lamentation of his life and family, Al comments, "If only they'd had a Scared Single program when I was a punk!" His opening "birthday discussion" with son Bud further illustrates Al's disdain for his role in the Bundy family. At his wife Peg's prompting, Al begins to offer Bud some words of wisdom for his eighteenth birthday. "Eighteen years old, huh?" Al begins. "There's so much I want to say to you, but there's a show coming on I want to watch." A *Newsweek* article stresses Al's rejection of "good fatherhood" still further:

> Probably the best of the show's 200-plus episodes is a parody of "It's a Wonderful Life," with the late Sam Kinison as Al's guardian angel. In a reversal of the Jimmy Stewart scenario, Al sees how *happy* his family would have been had he never been born. He can't allow it, bellowing "I want to live!" so they can be reunited in misery once again. The scene is the exact opposite of what's cynically known among sitcom writers as the MOS, or "Moment of S—." That's the cloying denouement when a character experiences some maudlin epiphany that inevitably results in hugging. (Marin, 1996, p. 70)

Focusing on the ways in which Al violates conventions of sitcom fatherhood, this article stresses Al's departure from the conventionalized masculinity of its preceding epoch. Or, as another writer puts it, "It was against this saccharine, Reagan-era approach to sitcom construction that *MWC* rebelled" (Lusane, 1999, p. 14).

In these ways, Homer Simpson and Al Bundy seem to turn the masculinity of the '80s on its head. Moving away from the ultra-comfortable world of the '80s dad, in which goods are abundant and consumption equals love, Homer's and Al's is a fatherhood of parodic desires in which dads can barely remember their families, let alone support them. Here, Homer and Al do seem to have laid waste to the nuclear father of the '80s, as suggested by the show's celebrants above. With these '80s fathers, consumption was a subtext, the natural state of being of the beaming Reagan dad. In contrast, the hyperbolic, hypermasculine desires of Al and Homer make this consumption explicit, highlighting the politics of consumption that might otherwise have remained conspicuously unquestioned. Likewise, their willingness to belittle their children seems an explicit challenge to the "everything is fine" images served up by the '80s. While Al and Homer are not the first to take fatherhood so lightly or to offer outlandish parodies of hypermasculine fatherhood, theirs is a uniquely hyperbolic version of this parodic restatement.[4] In their explicit rejection of '80s fatherhood, Homer and Al twist masculinity in ways that resonate with the contemporary crisis explored throughout my investigations here. Coming to terms with criticisms of masculinity, Al and Homer seem to scoff at the Reagan TV dad, mocking the abundant comfort, consumables, love, and beaming smiles of this '80s masculinity seemingly content with its invisibility.

And yet, this parody inevitably falls back on the "working-class masculinity" represented by Al and Homer, holding middle-class masculinity and cultural capital at a safely ironic distance. While Homer and Al's hyperconsumptive, hyperbolic activities, *do* mount a radical parody, the parody itself depends upon a middle-class understanding and appreciation of "taste." Indeed, the parodic power of Homer and Al comes not only from their hyperconsumption but also from their violation of norms of middle-class taste. Like Bill Clinton, another "convivial bon vivant" (Bourdieu, 1984, p. 194), whose appetite for fried foods and donuts has been a source of humor and derision, Al and Homer's appetites seem to violate middle-class taste. From rare gummi bears to rare copies of *"Big Uns"* (Al treats his collection with an antique-like reverence), Homer and Al's "tastes" are framed as comical, as hyperbolic versions of some "lowbrow" sensibility.[5] In short, dads Homer and Al mock the '80s fathers, whose abundant smiles, sweaters, and comfort are now seen as representative of a delusional Reagan-fantasy akin to the hard-body discussed in chapter 3. However, they do so only by taking shelter in that same middle-class fantasy, relying on this dream of cultural capital for their parodic humor.

Like the infantile citizen discussed by Lauren Berlant (1997), so Homer and Al (and Beavis and Butthead) offer a consoling image that works to negotiate a part of the contemporary crisis of masculinity, protecting the unblushing male's middle-class status even as they offer it up for parody. As Berlant suggests:

> The overorganizing image or symbolic tableau emerges politically at certain points of structural crisis, helping to erase the complexities of aggregate national memory and to replace its inevitably rough edges with a magical and consoling way of thinking that can be collectively enunciated and easily manipulated, like a fetish…. This means that the politically invested overorganizing image is a kind or public paramnesia, a substitution for traumatic

loss or unrepresentable contradiction that marks its own contingency or fictiveness while also radiating the authority of insider knowledge that all euphemisms possess. (p. 48)

The defilement of '80s fatherhood suggested by Al and Homer is just such an overorganizing image, offering "consolation" through an illusory critique. In the end, the conventionalized likes of Cosby et al. still beam happily, but now between the lines of Homer and Al's supposed parodies.

Working-Class Experience and Middle-Class Taste

While dominant perceptions of working-classness lend themselves to parodic restatements, like those of Beavis, Butthead, Homer, and Al, working-class masculinity also maintains a sense of real-world experience that is indicative of certain traditions of masculinity. As noted above, President Clinton's perceived "redneck" background has allowed him to distance himself from the corrupt world of monied politics, positioning him as another kind of candidate, more down to earth and in touch with "the people." Benefiting from perceptions of the "Man in the Gray Flannel Suit," which have framed middle-class men as automatonic, unfeeling robots, the down-home masculinity Clinton seems to evidence is one of earthy passion and real-world experience.

This masculinity of "working-class experience" has an important history in mediated form as well. Biskind and Ehrenreich (1987) discuss the ways in which *Rocky I* and *II*, for instance, depict a romanticized vision of working-class life.

> *Rocky I* and *II* romanticize [the working-class world]. Rocky himself, a thirty-year-old, over-the-hill fighter, is a noble savage, a natural man—Truffaut's Wild Child plopped down in south Philadelphia. Images of nature abound. Rocky keeps two turtles, whimsically named Cuff and Link, and a goldfish, and in Rocky II he buys a dog. He meets his wife in a pet store and proposes to her in a zoo. Even the fact that Rocky's employer (in *Rocky I*) is a gangster and Rocky's job is collecting bad debts doesn't taint the idyll one bit. (p. 211)

Indeed, the Rocky films celebrate Rocky's tough, working-class experience, depicting him as a survivor who scraps his way to the top despite the pressing odds against him. In *Rocky I*, he manages to fight

Apollo Creed, an experienced professional fighter, to a draw. While Creed trains on state-of-the-art equipment, Rocky spars with a side of beef at a meat processing plant, evidence of Rocky's distance from the moneyed world of boxing success. Indeed, in *Rocky III*, which pits Rocky against the powerful Clover Lane (Mr. T), it is Rocky's new-found success and connection to the upper-class world that causes his defeat. Instead of his traditional style of training, the new, wealthy Rocky works out in a fancy gym amid a media and public relations frenzy, a virtual amusement park for his fans. Only after he returns to his old ways, doing one-handed push-ups in a dank old gym, is Rocky able to fight Lane and win. Throughout these films Rocky evidences a romanticized vision of working-class masculinity, depicting the struggling working-class survivor whose determination, hard work, and distance from middle- and upper-class comfort eventually pay off.[6]

The images of masculinity explored in this section illustrate a similar romanticization of working-class masculinity, but with a '90s twist. Fathers and sons Homer, Al, Beavis, and Butthead demonstrate a pervasive, hyperbolic irony that protects middle-class masculinity even as it seems to critique it, evidencing their own negotiation of these middle- and working-class tensions. In contrast, the working-class masculinity illustrated by Jack Dawson (Leonardo DiCaprio) in *Titanic*, a '90s "son," as well as by '90s "fathers" Justin Matisse (Harry Connick Jr.) from *Hope Floats* and George Malley (John Travolta) from *Phenomenon*, is more of this romanticized type. Just as Rocky is celebrated for his connection to the street, so these characters are celebrated for their own working-class wisdom and experience, presented as a commonsense, hands-on knowledge of the world. But while Rocky's romanticized working-class status allows him to ascend to heights of boxing success, the working-class experience of these '90s characters is insufficient in and of itself. While Jack Dawson, Justin Matisse, and George Malley all embody a sense of working-class capital, this capital is ultimately supplemented with a quality of middle-class cosmopolitanism and cultural capital that seeks to balance out this lower-class "experience." Hence, these characters colonize working-class masculinity in multiple ways, offering up a sort of working-class "noble savage," while simultaneously trying to rescue him from his savagery.

Jack Dawson: Betwixt and Between Classes on the Titanic

Jack Dawson, Leonardo DiCaprio's heroic character in *Titanic*, offers up a romanticized look at a working-class free spirit. An artist who wins his third-class tickets in a poker game, Jack seems an instinctive creature who roams wherever his luck and fancy lead him. "When you got nothing, you got nothing to lose," Jack says, as he places his final bet in the poker game. Following where fate and whimsy lead him, Jack is not unlike Rocky Balboa, his connection to the "real world" and his skills of survival determining his life from one moment to the next. Stressing this romantic vision, an article in the *New Yorker* claims:

> The hero of the tale is Jack Dawson (Leonardo DiCaprio), a freewheeling scruff—a skinny kindred spirit, perhaps, of his namesake Jack London—who wins a couple of steerage tickets in a poker game. Armed with little more than a sketchbook and an insolent grin, he leaps aboard the Titanic as she is about to cast off from Southampton on her maiden—and, as it turns out, her funeral—voyage. (Lane, 1997, p. 156)

Similarly, an article in *The Nation* calls Jack "a working-class artist and rover" (Pollitt, 1998, p. 9), while a *Vogue* writer depicts him as "a charmer possessed of an other-worldly charisma: He's got a manly man's ability to protect a woman, but the face of a beautiful, sensitive boy" (Powers, 1998, p. 78). Finally, a *Newsweek* article observes "the thing about Jack Dawson—and what made him difficult to play for a child of the '90s like DiCaprio—is that he doesn't have a dark side. DiCaprio had never played a character without demons" (Ansen, 1998, p. 62). Imagined as a working-class innocent, free to roam the world, Jack offers a romanticized, idyllic vision of "working-class experience."

Likewise, just as the acting histories of Patrick Swayze and Wesley Snipes, in the previous chapter, provide an intertextual meaningfulness that contributes to their characters' negotiations of contemporary masculinity, so DiCaprio's own history as an actor enhances this romanticized vision of working-class experience. Even in his earlier television appearances as character Luke Brower on ABC's *Growing Pains* DiCaprio was cast as a sort of working-class innocent. An adolescent abandoned by his truck-driver father, DiCaprio's character is taken in by the middle-class Seaver family, a working-class youth who is cleaned up by the Seavers' upper-middle-class capital. His roles as Arnie Grape, a mentally challenged member of a rural, lower-class family in *What's Eating Gilbert Grape*, and as Jim Carroll, a working-class youth struggling against heroin addiction in *The Basketball Diaries*, offer similar working-class images. In *The Basketball Diaries*, DiCaprio's character, Jim, like Luke from *Growing Pains*, is pulled out of his lower-class circumstances by achieving a newfound cultural capital, here by winning recognition as a successful writer. Finally, even DiCaprio's Romeo in Baz Lurmann's 1996 version of *Romeo and Juliet* depicts a sort of romanticized vision of the working-class. After fleeing Verona to avoid being arrested for murder, DiCaprio's modern-day Romeo hides out in a deserted trailer park, smoking cigarettes and driving around in a dirty, beat-up circa 1970s convertible. Depicting a tamed-down version of a working-class rebel, DiCaprio's intertextual history makes him well prepared for the romanticized working-class vision of *Titanic*.

Indeed, this romanticized vision of working-class experience is developed throughout *Titanic*. From the film's beginning, DiCaprio's character, Jack Dawson, is presented as a sort of noble savage at odds with the opulence and snobbery of the first-class passengers. As the *Titanic* is at port, ready to set sail for America, these first-class passengers dethrone their luxury automobiles, overcrowded with suitcases, trunks, hat boxes, and the host of other goods that will accompany them on their trip. Cal Hockley, the son of a Pittsburgh steel tycoon and Jack's evil first-class nemesis, tips a porter to handle the bags for his family as well as the family of his fiancée, Rose DeWitt Bukater (with whom Jack quickly falls in love). Dressed in their fancy suits and hats, these first-class passengers are the bastions of wealth and high society. Similarly, their first-class staterooms are decorated with teak and crystal and silver appointments, illustrating the luxury they seem to believe they deserve. In obvious contrast to this luxury, after winning his steerage tickets Jack runs to jump aboard the *Titanic*, his only luggage a worn duffel bag slung over his shoulder. Likewise, his sparse stateroom, a gray square lined with metal bunk beds, further highlights his distance from this first-class opulence. Walking on the third-class deck, one of Jack's fellow passengers, a working-class Irish man, notes "First-class dogs come down here to take a shit," to which Jack responds "It lets us know where we rank in the scheme of things." Traveling in the bowels of the ship, Jack appears the antithesis of this first-class comfort, an everyday hero attached to the experiences of the common man.

Jack's romantic working-class status is further demonstrated in his interactions with Rose, who quickly leaves her fiancé Cal and finds this third-class traveler. When Rose becomes fed up with Cal and frustrated with her own upper-class life, she decides to end her pain by jumping off the bow of the ship. However, Jack is there to stop her, both talking her out of jumping and rescuing her when she slips trying to climb to safety. Here, the film begins to evidence the romanticized commonsense wisdom it accords with Jack's supposed working-class status and connection to the common man. Calmly talking to her as she stands at the bow of the ship, Jack slowly, but effortlessly, talks Rose out of jumping. "Stay where you are. I mean it. I'll let go," Rose tells Jack as he nears her. "No, you won't," Jack replies calmly. "What do you mean 'No, I won't'?" Rose answers angrily. "Don't presume to tell me what I will and will not do. You don't even know me." Remaining calm, Jack answers Rose with a commonsensical tone: "Well, you would have done it already." Jack's commonsense assessment of the situation indeed seems to be right, as Rose gives Jack her hand and lets him help her back aboard the ship. Here, Jack illustrates a romanticized commonsense wisdom, a powerful everyday knowledge that seems to emanate from his experiences among the common folk.

After this initial encounter, Rose and Jack spend time getting to know each other, another opportunity to discover Jack's romanticized working-class experience and everyday wisdom. Walking together on deck, Jack tells Rose about his rough childhood in Wisconsin. "I've been on my own since I was fifteen, since my folks died," Jack begins. "I had no brothers or sisters or close kin in that part of the country. So I lit out of there and haven't been back since. I guess you could just call me a tumbleweed blowin' in the wind." This picture of Jack's "freedom" proves an important component of his romanticized working-class experience and wisdom and is further developed through discussions of Jack as an artist as well. Looking through Jack's sketchbook, Rose is surprised by the quality of his sketches. "What are you, an artist or something?" she asks, looking through the book's pages. "These are rather good. They're very good, actually," she says with a surprised tone; "Jack, this is exquisite work." "They didn't think to much of it in gay Paris," Jack answers. "Paris?" Rose responds with a like tone or surprise. "You do get around for a po ... a person of limited means." Being sure to stress his lower-class status, Jack emphasizes "A poor guy. You can say it." Here, Jack's lower-class status is conflated with his freedom and his commonsense wisdom, bringing this romanticized ideal into clear relief. Looking at these sketches ("drawn from life," Rose highlights), Rose compliments him. "You have a gift, Jack, you do. You see people." From working on a squid boat to doing portraits on the Santa Monica pier (as Jack explains to Rose as they talk), Jack seems to float through life, collecting knowledge and wisdom to see the world, and people, for what they really are.

This romanticism is still more clearly evident in a scene in which Jack has dinner with Rose, Cal, and their wealthy friends and family. Borrowing a tuxedo from Molly Brown, a woman whose status as "new money" keeps her at a similar distance from these upper-class socialites, Jack does his best to fit in with this upscale crowd. "You shine up like a new penny," Molly Brown tells him as she looks at him in his tux. Dressed in these borrowed clothes, Jack tries to imitate the walk and mannerisms of the wealthy passengers who pass him in the fancy, formal dining hall. Holding his arm at his side, he mimics the way one elderly gentlemen shakes hands, attempting to look as upscale as possible. When Rose enters, Jack kisses her hand, commenting "I saw that in a nickelodeon once and I always wanted to do it," his reference to nickelodeons further evidence of his working-class roots. "Amazing," Cal comments. "You could almost pass for a gentleman." Indeed, the uneasiness with which Jack executes his gentlemanly performance serves to reinforce his romantic working-class experience. Struggling with his napkin or his silverware, Jack looks uneasy with these upper-class comforts, proof, no doubt of his clear tie to working- and lower-class sorts of experiences.

"Tell us of the accommodations in steerage, Mr. Dawson," asks Rose's mother, Ruth, apparently uneasy that her daughter has taken up with a working-class youth and hoping to highlight his lower-class status. "The best I've seen ma'am, hardly any rats," Jack answers jokingly, evoking a chuckle from the rest of the dinner company and turning this snobbish jab into a demonstration of his own cleverness. "And where exactly do you live?" Ruth follows up, attempting again to highlight Jack's inferior status. "Right now my address is the *RMS Titanic*. After that I'm on God's good humor." Feigning interest, Ruth asks Jack to explain how he manages to travel, given his limited means. "I work my way from place to place, you know, tramp steamers and such," Jack explains. When Ruth asks how someone could find such a "rootless existence appealing," Jack launches into a sort of vagabond's manifesto, a long explication of his romantic world vision:

> I got everything I need right here with me. I've got the air in my lungs and a few blank sheets of paper. I mean, I love waking up in the morning not knowing what's gonna happen or who I'm gonna meet, where I'm gonna wind up. Just the other night I was sleeping under a bridge and now here I am on the grandest ship in the world having champagne with you fine people.... I figure life's a gift and I don't intend on wasting it. You never know what hand you're gonna get dealt next.... You learn to take life as it comes at you, to make each day count.

Summing up this romanticized vision clearly, Jack's confident speech lectures these upper-class travelers on the joys of freedom and the pleasures of a "rootless," lower-class existence. Evoking a toast "to making it count" from this band of wealthy passengers, Jack's commonsense wisdom seems to win out over their educated, upper-class sensibilities.

While Jack's lower-class experiences are romanticized, however, the other lower-class passengers on Titanic do not fare as well, proving foils against which to develop Jack's romantic sort of working-class experience. For instance, when we first see the *Titanic* in port, an apparently lower-class onlooker holding a young child looks at its massive frame. "Big boat, huh?" he comments to his young daughter. "Daddy, it's a ship," she corrects him. Needing correction from his younger daughter, this particular lower-class character lacks the sort of wisdom accorded to Jack's working-class experience. Similarly, after the dinner party with Rose and her wealthy friends and family, Jack takes Rose away to a "real party" being put on by the steerage passengers of the third class. This raucous party below deck is a far cry from the staid and proper dinner atmosphere in which Rose and the other first-class passengers take part. While this scene demonstrates a sort of romanticized expression of working-class joviality, in which these passengers throw aside staid propriety and immerse themselves in a good time, this romanticism is different from that developed with Jack's character. Falling around the room in drunken chaos, spilling beer, and collapsing on the floor, these steerage passengers are distanced not only from the upper-class passengers above deck but from Jack as well. Indeed, when one drunken man spills beer on Rose, Jack chastises him to clean up his act. Whereas Jack can "clean up," traveling however uncomfortably to the upper-class world of the upper decks, these other passengers seem permanently stuck below decks.

In this world of first class and steerage, Jack is thus presented as *both* and *neither*, his romanticized version of working-class experience granting him a cosmopolitanism that ensures a certain middle- or upper-class cultural capital while maintaining a connection with "everyday folk." His travels around the world have seemingly afforded him this place betwixt and between classes. His way of speaking, for instance, seems neither lower nor upper class, distanced both from the people in steerage as well as from those in the upper decks. While he lacks the snobbishly affected accent of Cal and friends, the heavily cockney or working-class Irish accents of the other steerage passengers signal their lower-class status more obviously than Jack's relatively nondescript Midwestern accent ("You'd as likely have angels fly out of yer arse as get next to the likes of her," an Irish friend comments when he sees Jack pining after Rose). Likewise, while Jack can get tough—after the ship sinks, he helps rescue Rose by punching out a man holding on to her—his status as artist keeps him separate from the muscle-bound, coal-covered crewmembers in the boiler room. When Jack inadvertently stumbles upon the boiler room, one worker sternly warns, "You shouldn't be down here, you could get hurt," suggesting Jack's distance from this sort of working-class environment. In contrast to this world of manual work, Jack's travels around the world have brought him experiences approaching Rose's upper-class education. "Monet," Jack exclaims excitedly, correctly identifying a painting in Rose's stateroom. Able to dance a jig and drink pints of beer with the folks in steerage yet still identify great art work, Jack's working-class experience bears a decidedly middle- or upper-class cultural capital. Negotiating these senses of working-class masculinity, Jack borrows from working-class experience but doesn't succumb to it, keeping it at arm's length even as he appears to romanticize it.

Offering this sort of negotiated working-class experience, the message of *Titanic* vis-à-vis class and masculinity seems fairly clear. Although working ideals can help reframe popular notions of masculinity amid crisis-of-masculinity talk, this experience must be of a certain sort. Maintaining working-class characteristics for some romantic connotation of folksiness (the sort that has helped Clinton), Jack's character colonizes a stereotypical working class of joviality and indulgence. And yet, this same stereotypical working-class masculinity is also held at bay, pushed aside for its connotations of slavishness and its rejections of the cultural capital associated with the unblushing male. In the end, the romanticized noble savagery of Jack Dawson is similar to, and different from, that of the earlier figure Rocky. Whereas Rocky was heroic because of his tough working-class background, Jack seems heroic in spite of his, overcoming this lower-class masculinity through the accumulation of a certain kind of cultural capital. "I'm a survivor," Jack tells Rose as he puts her on a lifeboat, attempting to reassure her that he will make it through *Titanic*'s sinking. Though he ultimately drowns, Jack Dawson's ability to negotiate these contemporary conceptions of masculinity seems a strong survival skill indeed.

In these ways, Leonardo DiCaprio's Jack Dawson negotiates the contemporary climate of masculinity differently from fellow '90s sons Beavis and Butthead. All lacking fathers, Beavis, Butthead, and Jack Dawson seem evocations of a '90s welfare child—adolescents left free to make their way through contemporary culture. Whereas Beavis and Butthead offer a sort of parody of working-class youth, however, Jack Dawson seems to embrace the possibilities of lower-class youthfulness, celebrating its freedom and possibilities. Yet, this supposed possibility is ultimately subverted, blunted by Jack's romanticized vision of working-class manhood, his accumulation of cultural capital and his juxtaposition to the "bad working class" of the film. Thus, although Beavis and Butthead's image of working-class life is read as parodic and Jack Dawson's as heroic, they both afford the same renunciation of working-classness. For Jack, his accumulation of cultural capital proves the antidote for his "working-class experience," pushing it aside even as it is apparently embraced.

Childless Fathers: John Travolta and Harry Connick Jr.

If Jack Dawson is the '90s heroic image of a lower-class son, his lower-class status subverted by his accumulated cultural capital, then Justin Matisse and George Malley may be its heroic welfare fathers. In contrast to Homer Simpson and Al Bundy (working-class dads who shun fatherhood and put their own needs above their children's), Justin and George quest to *become* fathers. Indeed, both Justin and George fall in love with single mothers, eventually convincing them that they can provide the fatherly love their children are missing. In the characters of Homer and Al, working-class status becomes an excuse for hyperconsumption, a sort of hyperbolic parody of the '80s dad that ultimately reels back upon these working-class men. Despite their own working-class connections, however, both Justin and George put family first, running towards, rather than away from, a commitment to family. Ultimately, however, this commitment is not enough; Justin and George are granted a middle-class cultural capital that, similarly to Jack Dawson's, holds their working-class masculinity at bay. Romantically working class, yet endowed with middle-class cultural capital, Justin and George are the model new welfare dads— committed to family and endowed with the "right cultural values" to make these families work.

Justin Matisse, Harry Connick Jr.'s character in *Hope Floats*, works as a house painter and carpenter, doing odd jobs to make ends meet. A stereotypical Texas work hand, Justin wears cowboy boots, blue jeans, and a straw cowboy hat. Attempting to woo Birdee Pruitt (Sandra Bullock), a recently divorced single mother who has returned home to live with her parents, Justin makes frequent visits to her home, prompted, in part, by Birdee's mother's attempts at matchmaking. In one early scene, Birdee's mother, Ramona, has invited Justin over to dinner, hoping to strike a spark between Justin and the recently returned Birdee. "Are you wearing cologne?" Ramona asks Justin, trying to get Birdee's attention. "No, that's a little paint thinner, a little sweat mixed in with it," Justin answers, stressing his everyday folksiness. Indeed, similarly to Jack Dawson, Justin maintains a folksy sensibility and wisdom that romanticizes his working-class experience. When he sees Birdee eating alone at a local diner, Justin, himself alone at a table, offers her the benefits of this folksy wisdom: "It's not for sissies, you know, dining alone," he explains, "the trick is to seem mysterious, like the choice is yours." Likewise, when Birdee is nervous about dancing, Justin explains to her sagely, "Dancin's just a conversation between two people. Talk to me." Further emphasizing his folksiness, after talking Birdee into going out for a "nice fish dinner," Justin takes Birdee fishing, frying the fish they catch over an open fire. Dressed in his "Texas Rebel Radio" T-shirt and cowboy hat,

Justin seems a romantic picture of working-class masculinity, a Marl-boro-man-like rebel whose everyday wisdom helps him enjoy people, and the world, for what they are.

George Malley, John Travolta's character in *Phenomenon*, demon-strates a similarly romanticized vision of working-class life. Set in a small town in northern California, the film's opening shots of the sun rising on cornfields, broken-down pickups filled with straw, wind-mills, and stray kittens climbing on old tires set the stage for this idyl-lic depiction. A tow-truck driver and auto mechanic, George dresses in blue jeans when not in his dark blue mechanic's coveralls. His old house, a white farmhouse with the paint beginning to peel and the picket fence beginning to pitch over, furthers this stereotypical pic-ture of small-town life, as do his chicken-wire-enclosed vegetable garden and his old hound dog, Atilla. At home with his friends, George fixes cars by day and drinks beer by night, making visits to his local small-town bar. When a friend gives George some high-tech solar panels to install on his house, George observes, "If it's a car, I can fix it. If it's in the garden, I can probably grow it. But this is way out of my neighborhood." Connected with the everyday sorts of ex-periences of Justin Matisse and Jack Dawson, George is a regular guy, happily struggling his way through his life.

At the same time that Justin's and George's characters seem to romanticize working-class life, their characters, similar to Jack Daw-son, also negotiate a sort of bourgeois cultural capital. While Justin employs himself by painting houses, for instance, his real passion is the great architecture he secretly designs and builds on his property. This work of great genius is introduced when Justin takes Birdee for a tour. "Okay, be kind," Justin tells her as they prepare to enter. "It's a work in progress." Justin opens the door, turns on the light, and un-veils an architectural and interior design masterpiece. Its walls made from "nineteenth-century Texas pine" (as Justin explains) but the an-gles and shape giving a modern feel, the house seems a sophisticated marriage of traditional and modern. The stained-glass ceiling, for in-stance, looks like a combination Aztec painting and Frank Lloyd Wright design. "The porch. It's gonna have a great view once I get rid of the trailer," Justin explains, pointing towards a dilapidated trailer in the backyard. "Who was your architect?" Birdee asks. "You're kid-din' huh?" Justin replies. "That's half the fun." "You go around Smithville painting houses and you can do this?" Birdee observes, still in awe of the house, "You could do so much more." In his own manifesto of freedom (reminiscent of Jack Dawson's), Justin responds:

> [You're] talkin' 'bout the American dream. Find something that you love and then you twist it and you torture it. Try to find a way to make money at it. Spend a lifetime doing that. And at the end you can't find a trace of what you started out lovin'.

Capable of designing great architecture, Justin chooses to live like Thoreau ("I came back here so I could live the way I wanted," Justin explains, as if quoting from *Walden*). Maintaining this bourgeois taste and ability in architecture, Justin nonetheless keeps it secret, negotiating his choices in a way that maintains his everyday working-class experience as well as his bourgeois cultural capital.

Whereas Justin is a gifted architect, George Malley's developing cultural capital comes about more haphazardly. Blinded by a strange light in the sky, George suddenly develops a thirst for knowledge and an ability to master almost anything he pursues. While the earlier George could not install a solar panel, this newly gifted George does this and more, conducting experiments with solar energy and explaining to a friend that "the whole field of photovoltaics is in babyland." Likewise, George learns Spanish overnight and most of the Portuguese language in twenty minutes. He engineers a new fertilizer for his garden that allows him to grow the largest tomatoes the county has seen. He builds a methane-powered engine driven by his friend's pig manure and his own garbage. He even develops telekinetic powers that allow him to move things at will. "Lately I've been seeing things so clearly." George explains. "Books—I read two to three books a day. Imagine that. I can't stop. I can't sleep." Apparently tapping into the multiple possibilities of his brain, George demonstrates the many things of which the human mind is capable. Granted these new powers, the everyday George Malley has become an extraordinary conveyor of scientific knowledge.

Like Justin, however, George is somewhat uncomfortable with his own abilities. Frustrated with his ability to handle his newfound intelligence, at one point in the film George yells, "It was a goddamn mistake. It was supposed to happen to someone smart, someone scientific, someone who was a leader. But it didn't. It happened to me, George freakin' Malley." Unable to sleep, George is overwhelmed by his new intellectual energy, struggling to find ways to occupy his now busy mind. When a professor from Berkeley comes to visit, hoping to learn how George was able to predict an earthquake before it happened, George prods him for a visit to the campus and an audience for his various inventions. "I really need to talk to people like you," George explains to the professor desperately. Alone in his working-class world, George feels like a freak, a material laborer suddenly thrown into the world of mental work.[7] And yet, maintaining a sort of negotiation, George eventually comes to terms with his new intellect, though only when he learns that it will soon kill him. After doctors discover that a brain tumor has been stimulating activity throughout his brain, George is admitted to the hospital, the doctors hoping to research his brain activity—to do a sort of "expedition" into this phenomenal new mind. At this point, George delivers his own treatise, summing up the incredible potentials locked away in all common folk:

> I think I'm what everybody can be…. Anybody can get here. I'm the possibility, all right? What I'm talking about is the human spirit. That's the challenge. That's the voyage. That's the expedition.

George is not phenomenal at all, he seems to say, but only evidence of what everyone can become if they apply themselves. Of course, George's difficulties send a mixed message. Certainly great brainpower may be inside everyone, but its exercise seems to take a greater toll on some people (i.e., the working class) than others. Indeed, George's mental activity eventually kills him.

While Justin and George struggle with their own cultural capital, they also struggle to become fathers, to endear themselves to a single mother and her children. Justin, for instance, must win over Birdee, her daughter Bernice, and her nephew turned foster child. Bernice, like Birdee, is initially skeptical of Justin. When Justin comes to their house Bernice mocks him, poking fun at his clothing, his way of speaking, and a bouquet of flowers he brings. Hoping that Birdee will reunite with her father, Bernice holds out a dream of a typical American family, sure that her daddy will take her away from her frustrations in Smithville. Still, Justin maintains a constant presence, demonstrating to Bernice, Birdee, and Travis that he is worthy of their love. After Birdee's mother, Ramona, dies, Justin comes over to visit, spending time on the porch with Travis. Sitting together on a porch swing, Justin plays sensitive father-figure. "My grandma's dead," Travis says sullenly. "I know," Justin answers, sitting quietly beside him. "Aren't you going inside?" Travis asks after a few minutes of waiting together. "Don't you know?" Justin answers. "I came over to say hi to you." "Being there for Travis," Justin seemingly demonstrates his fatherly affection and potential. Indeed, the film's closing scene has Justin, Birdee, Bernice, and Travis walking hand-in-hand towards their house—one big happy family finally united in love and harmony.

In the same way, the phenomenal George Malley eventually proves himself a worthy father figure as well, *Phenomenon* offering up a similarly happy family conclusion. Early in the film, George struggles in the same ways as Justin, not immediately presenting himself to Lace, his single-mother love interest, as worthy of her kids. Driving out to visit their house, George offers a ride to Lace's children, Al and Glory, asking them to help him pick flowers as a present for their mother. When he gets to their house, he presents the flowers to Lace, explaining that the kids helped select them. "I've got to tell you," he says to Lace, "you've got great kids." "Yeah, they are great kids," Lace answers, "but you don't really know that yet, George, because you don't know them and you don't know me." Wanting to keep her life simple, Lace explains that she doesn't want to get involved with George or anyone else. Of course, the development of the film seems to change Lace's mind. Seeing George's struggles, she eventually begins to care for him, helping him negotiate the problems he experiences from his new mental powers. Indeed, by the film's end, with his inevitable death looming, George moves in with Lace and the kids, becoming the same big happy family presented at the end of *Hope Floats*. Playing his fatherly role, a concluding scene has George, Al, and Glory standing together, talking against a wire fence, George helping the kids deal with his impending death:

> If we were to put this apple down and leave it, it would be spoiled and gone in a few days. But if we were to take a bite of it like this, it would become part of us and we could take it with us forever.

Delivering this folksy wisdom, George gets Al and Glory to take bites of the apple, teaching them an important lesson about life, death, and memory. Now a happy father, George passes away that evening, tucked into bed in his new house, amid his new family.

Both Justin and George thus stand as powerful complements and contrasts to fellow '90s dads Al Bundy and Homer Simpson. Al and Homer parodically reclaim the Reaganomic middle-class legacy of the '80s. Presented as dysfunctional, working-class dads who reject their place as fathers, they put their own desires above the needs of their children. Justin and George pursue fatherhood, two working-class men drawn to the responsibilities that Al and Homer shun. Couched in their romanticized working-class existence, Justin and George are thus presented as the model welfare dads of the '90s, as the perfect evocation of a new working-class father. Colonizing a working-class noble savagery, Justin and George are endowed with a middle-class cultural capital that—as with Jack Dawson—holds their working-class status at bay even as it is purportedly romanticized. The heroism of both Justin and George, and thus their legitimate claim to fatherhood, seems to follow precisely from their hidden, middle-class potentials. George's speech about the human spirit attests, as do both of these films, that middle-class abilities and sensibilities are buried deep inside every contemporary male. Those who have harnessed them, it seems, are heroic, like George and Justin, while those who haven't are parodic, like Homer and Al. Like the self-made man of American history, these working-class men are self-made (or failed) dads—their middle-class sensibilities direct reflections of their earnestness in harnessing their own potentials.

National Dads and National Children

So these '90s men work to negotiate the masculine problematics of working-class experience and middle-class cultural capital. Simultaneously mocking and embracing the Reaganomic fantasy of the '80s, as do Homer, Al, Beavis, and Butthead, and/or romanticizing and rejecting working-class experience, as do Jack, Justin, and George, these popular deployments demonstrate an anxious attempt to come to terms with class vis-à-vis masculinity. In *Policing the Crisis*, a discussion of mugging, a sort of criminal activity associated with the lower classes, Stuart Hall et al. demonstrate a similar sort of negotiation. They write, "its treatment evokes threats to, but also reaffirms the consensual morality of the society: a modern morality play takes place before us in which the 'devil' is both symbolically and physically cast out from the society by its guardians" (1978, p. 66).[8] The sort of "morality play" illustrated here is similar and different. Rather than "casting out a devil," the '90s fathers and sons illustrated here take up a version of working-class masculinity, only to reinscribe it in a middle-class fantasy that recalls the likes of Goffman's unblushing male.

Of course, these same tensions are busily at play in the Clinton persona as well, evoked within the redneck/policy wonk struggle noted at the opening of this chapter. Negotiating tensions of working-class Southernness with those of Oxford-educated Rhodes Scholar, Clinton sits precariously between experienced man of the people and backwoods clod. Just as these men sit distanced from and embracing middle-class fatherhood, so Clinton appears both the parody of, and ideal model for, national daddyness. In the midst of the Monica Lewinsky scandals, for instance, Clinton opened up a fatherly parody not unlike Homer and Al. As the *Washington Post* explains it, "An editorial cartoon in the relatively staid *Birmingham Post* captures the mood. It shows a young woman in a miniskirt heading out to a party. Her father stops her and says, 'You're not leaving this house, young lady, as long as that Bill Clinton is in town'" (Reid, 1998, p. 30). Here, Clinton's threat to daughters reflects back upon his own fatherhood. A daddy with a hyperbolic sexual appetite, Clinton seems not unlike Al Bundy with his stacks of "*Big Uns*" and trips to the nudie bar.

At the same time, however, the *New Republic* suggests that it is Clinton's positioning as "national Dad" that "accounts for the rise in [his] approval ratings since the accusations concerning Monica Lewinsky were first made" (Rubin, 1998, p. 14). As the article explains,

> The family paradigm has played an important role in the politics of this presidency. In a memo to Clinton advisor Dick Morris early in the 1996 campaign, feminist author Naomi Campbell urged Morris to portray Clinton as the "Good Father" protecting the family home from Republicans intent on destroying it. Morris kindled to the idea. As he later recounted in his memoir of the 1996 campaign, *Behind the Oval Office*: "Arkansas saw Clinton as their son.... In the '92 campaign, Clinton was America's buddy.... But now, I told the president, it's time to be almost the nation's father, to speak as the father of the country, not as a peer and certainly not as a child."

Recounting this story, this article understands Clinton's popularity precisely as his ability to be this national father. "This approach worked," the piece continues,

> precisely because the notion of Clinton as first father was not mere image manipulation—despite the involvement of spin doctor Morris in its formulation. The concept was not perfectly apt because of Clinton's youthful age and demeanor. But the idea of Clinton as metaphorical family member resonates because of who he actually is: a genuinely empathetic person who is most passionate about policies he believes will help save America's families.

Clinton's ability to carry off national fatherhood, while being offered up as a parody of hyperbolic consumption, demonstrates his bizarre negotiation of contemporary masculinities.

The ideas about fathers and sons illustrated in these '90s new men evidence a strange attempt to move among and between middle- and working-class stereotypes of manhood. Whether "parodying" a Reaganomic dream or "romanticizing" working-class maleness, these popular deployments fall back upon a middle-class cultural capital, pulling the unblushing male up by his bootstraps. In so doing, these '90s images illustrate the emotional meanings underlying this cultural capital—anxiously maintaining a safe sense of middle-class taste that champions middle expistence in the face of this new, presumably working-class male. Discussing such moments of struggle, Bourdieu argues that "principles of division, inextricably logical and sociological, function within and for the purposes of the struggle between social groups; in producing concepts, they produce groups, the very groups which produce the principles and the groups against which they are produced" (1984, p. 479). Putting this another way, Bourdieu emphasizes:

> The classifying subjects who classify the properties and practices of others, or their own, are also classifiable objects which classify themselves (in the eyes of others) by appropriating practices and properties that are already classified (as vulgar or distinguished, high or low, heavy or light, etc.—in other words, in the last analysis, as popular or bourgeois) according to their probable distribution between groups that are themselves classified. (p. 482)

In offering up these new male images, the above popular images reclassify a new middle-class masculinity, but not without classifying the working-class foil against which this new maleness is exercised.

Notes

1.   Barbara Ehrenreich's book *The Hearts of Men: American Dreams and the Flight from Commitment* offers a very insightful discussion of *Playboy* and '50s masculinity. While Ehrenreich stresses *Playboy* as a source of masculine escape (as does Kimmel, 1996, pp. 254–257), she also hints as its role in perpetuating a masculinity built upon consumption: "*Playboy* had much more to offer the 'enslaved sex' than rhetoric: It also proposed an alternative way of life that became ever more concrete and vivid as the years went on. At first there were only the Playmates in the centerfold to suggest what awaited the liberated male, but a wealth of other consumer items soon followed. Throughout the late fifties, the magazine fattened on advertisements for imported liquor, stereo sets, men's colognes, luxury cars and fine clothes" (p. 49).

2.   Pierre Bourdieu (1984) demonstrates the connection between "convivial indulgence" and ideas of working- and lower-class culture. "The art of eating and drinking remains one of the few areas in which the working classes explicitly challenge the legitimate art of living. In the face of the new ethic of sobriety for the sake of slimness, which is most recognized at the highest levels of the social hierarchy, peasants and especially industrial workers maintain an ethic of convivial indulgence" (p. 179). The way in which Clinton's excessive love of food is highlighted in public discourses contributes to his problematic and conflicted class status vis-à-vis masculinity.

3.   For a discussion of Fox's attempt to capture this youth market, especially as related to *Married … with Children*, see Lusane (1999).

4.   While Fred Sanford, of *Sanford and Son*, and Archie Bunker, of *All in the Family*, are similarly parodic, their different statuses as fathers prove important contrasts to Al and Homer's particular parodies. For instance, both Fred Sanford and Archie Bunker *have adult children*, presented as capable of standing up to, and resisting, their ridiculous antics (indeed, on both of these shows the children are noticeably different from their parents). So while they are certainly parodic, the parody does not suggest the same rejection of fatherhood. In fact, we can recall that Archie Bunker becomes significantly more subdued when presented as a father of a young child—when wife, Edith's, niece, "Stephanie Mills" comes to live with them in the last few seasons of the show.

5.   Bourdieu (1984) offers a way of understanding this sort of reproduction of middle-class capital. He writes, "The classificatory system as a principle of logical and political division only exists and functions because it reproduces, in a transfigured form, in the symbolic logic of differential gaps, i.e., of discontinuity, the generally gradual and continuous differences which structure the established order" (p. 480). The ways in which dominant groups manipulate components of cultural capital stereotypically associated with one or another class help to re-

produce these "differential gaps," defining and protecting middle-class capital, in this case, by this parodic representation of "lowbrow" taste.

6. Rocky's career seems a movement from lower- to upper-class culture and back again. When he has attained wealth, in *Rocky III*, his thick "working-class accent" seems mysteriously to have disappeared. In contrast, in *Rocky V*, set after Rocky has gone bankrupt and returned to his old neighborhood, his thick slang dialect has returned—now because of the brain damage he has sustained throughout his boxing career. All of this seems clearly to suggest the presumed connection between cultural capital and economic capital—Rocky's loss of economic capital necessitating the loss of his cultural capital as well.

7. In *The German Ideology*, Marx and Engels suggest that the division of labor "manifest itself also in the ruling class as the division of mental and material labour, so that inside this class one part appears as the thinkers of the class (its active, conceptive ideologists, who make the perfecting of the illusion of the class about itself their chief source of livelihood), while the others' attitude to these ideas and illusions is more passive and receptive, because they are in reality the active members of this class and have less time to make up illusions and ideas about themselves" (Tucker, 1978, p. 173). *Phenomenon* indeed seems to suggest that George is struggling with these ideas, a material laborer frustratingly dealing with mental labors.

8. For similar discussions within British cultural studies of this process of containing, reproducing, and/or casting out a vision of working-class masculinity, see Hebdige (1979), Willis (1977), and Hoggart (1957).

Chapter 4

The Exotic White Other:

Otherworldly Whiteness from Clinton to Fox Mulder

In October of 1998, well into the scrutiny of President Clinton's affair with Monica Lewinsky, Toni Morrison offered an interesting commentary for the *New Yorker*. Discussing Clinton, Morrison argues that:

> Years ago, in the middle of the Whitewater investigation, one heard the first murmurs: white skin not withstanding, this is our first black President. Blacker than any actual black person who could ever be elected in our children's lifetime. After all, Clinton displays almost every trope of blackness: single-parent household, born poor, working-class, saxophone-playing, McDonald's-and-junk-food-loving boy from Arkansas. And when virtually all the African-American Clinton appointees began, one by one, to disappear, when the President's body, his privacy, his unpoliced sexuality became the focus of the persecution, when he was metaphorically seized and body-searched, who could gainsay these black men who knew whereof they spoke? ("Talk of the Town," 1998, p. 32)

Marking Clinton as black, Morrison made problematic the manner in which Clinton's post-Lewinsky body had come under scrutiny. According to Morrison, the "message" writ upon Clinton's body was clear: "No matter how smart you are, how hard you work, how much coin you earn for us, we will put you in your place or put you out of the place you have somehow, albeit with our permission, achieved" — the *you* here, of course, representing bodies that are black, like Clinton's.

Clinton's potential blackness is an interesting problem through which to think about the contemporary crisis of masculinity.[1] That

Clinton can be "the first black President" further evidences the conflicted sorts of marking practices discussed throughout the preceding
chapters. Indeed, that Clinton can be read as black—not only in Morrison's performative reading here, but in the Clinton campaign's attempt to identify him with the black community—calls attention to
the cultural constructedness of blackness and, by implication, that of
whiteness as well. That this "physiognomically white" President
could be marked as "black" suggests the mobile and thus unstable
contours of whiteness, a potential instability important to much contemporary work within whiteness studies. As Richard Dyer suggests,
just as masculinity has stereotypically been seen as a universal, non-
gendered abstraction, so "white" has been framed as an abstract, non-
raced race. Says Dyer,

> The sense of whites as non-raced is most evident in the absence of reference
> to whiteness in the habitual speech and writing of white people in the west.
> We (whites) will speak of, say, the blackness or Chineseness of friends,
> neighbors, colleagues, customers, or clients, and it may be in the most genu-
> inely friendly and accepting manner, but we don't mention the whiteness of
> the white people we know. (1997, p. 2)

That Clinton may not really be white begs the question of what
whiteness *really is*, potentially forcing open this otherwise closed subject.[2]

Clinton's strangely racialized status demonstrates interesting and
important conflicts of identity. As Morrison suggests, Clinton's mobile and conflicted whiteness was an important component of his
presidency, making news long before stories of Monica Lewinsky
spread through the media. In October of 1992, an article in the *Washington Post* observed that "the rib joint and the country club are
geocultural icons for two separate worlds in Little Rock, one black
and the other white," noting that "Clinton seemingly has moved between the worlds with ease and regularity." Noting Clinton's movement between these worlds, this article calls him "a bicultural
explorer with redneck roots, a combination that provides him with
stronger empathetic ties to both worlds than most politicians, but also
leaves him open to apparent contradictions" (Maraniss, 1992, p. 1).
Similarly, according to a September 1992 article in the *New York Times*,

> More blacks and women have been appointed to boards, commissions and
> departments under Mr. Clinton than under any Arkansas chief executive.

Mr. Clinton has appointed blacks as the state's chief financial officer, health commissioner and head of the Department of Social Services. Yet, Mr. Clinton's achievements are no better than those of many other governors nationwide, and his administration has been criticized for its record in providing legal protection for blacks and improving their economic conditions. (Kolbert, 1992, p. 1)

In these ways, Clinton's "racial status," in terms of both his own ability to "pass" between black and white worlds and his position on important racial issues is seen as mobile, mutable, and conflicted.

Clinton's mobile whiteness has important implications for the discussions laid out here. Certainly, whiteness is an important component of Goffman's unblushing male in America and thus an important part of his invisible masculinity. Emphasizing the power of this invisible whiteness, Dyer maintains that one's ability to speak for some universalized humanity constitutes a particularly important cultural power. "Raced people can't do that—they can only speak for their race. But non-raced people can, for they do not represent the interests of a race" (1997, p. 2). Clinton's marked racial status, however conflicted, offers a potential challenge to this universality, suggesting the racialized status of the whiteness Clinton both exhibits and escapes. And Clinton is not alone in problematizing whiteness for the '90s, a decade that signaled a profound attention to racial issues and identities. The 1990 film *Dances with Wolves*, for example, attempts to reconcile the horrors of white history by creating a white character, John Dunbar (Kevin Costner), who self-identifies as Native American. "See," the film seems to say, "we weren't all bad," offering white America a hero through which to salve this horrible past. At the same time, the public arguments over the Rodney King beatings as well as the O. J. Simpson verdict dramatically highlighted the further problems of white America, demonstrating that racism and racial tensions are alive and well in the United States. Such moments offer critiques of whiteness (in some cases poorly handled, e.g., on the white shoulders of Kevin Costner), contributing to an importantly racial climate through which America—and American masculinity—has been forced to negotiate.

In addition to potentially disrupting the universality of whiteness—both by showing it as a construction and by forcing open its painful history—the conflicted racial identity of the '90s also participated in another conflict traditional to these conventionalized notions of masculinity. Just as class has a troubled relationship to dominant notions of masculinity, as argued in the previous chapter, so race relates to these dominant notions in equally problematic ways. As Kobena Mercer (1994) explains it, black men, for example, are "implicated in the same landscape of stereotypes which is dominated and organized around the needs, demands, and desires of white males. Blacks 'fit' into this terrain by being confined to a narrow repertoire of 'types'—the supersexual stud and the sexual 'savage' on the one hand, or the delicate, fragile and exotic 'oriental' on the other" (p. 133). As Mercer here suggests, nonwhite males are conventionally seen as both hypermasculine and hypersensitive or effeminate. The stereotype of the warring Native American savage exists alongside that of the natural spirit in sensitive touch with Mother Earth, as *Dances with Wolves* suggests in its depiction of the evil warring Pawnee and the good, buffalo-tracking Sioux.[3] Similarly, Asian men are conventionally represented as both "fragile," "oriental" others, and as vicious martial arts killers, the stealthy, deadly Vietcong soldier and passive Vietnamese villager both staple characters in countless war films.

Together, the conflicted, marked status of race in the '90s and the already conflicted status of race vis-à-vis masculinity created a series of important racialized negotiations for the '90s new male. This chapter explores these negotiations, investigating the ways in which a set of '90s males relates to, problematizes, and rehabilitates their whiteness. From Andy Kaufman (as portrayed by Jim Carrey) in *Man on the Moon* to *The X-Files'* Fox Mulder, these '90s men have increasingly mobile identities, allowing them to disrupt and distance themselves from whiteness at the same time that they espouse the necessity of its universality. On the one hand, these images are driven by a need to make amends, attempting to right the wrongs of whiteness and its history. Like John Dunbar, they find themselves, at times, identifying with "the other." Allying themselves with the others against whom white abstraction has practiced its dominance, they distance themselves from whiteness, holding up its horrific past. At the same time, however, they also work to mend whiteness, to stitch back together some of its universal nonidentity. New '90s men Andy Kaufman, Fox Mulder, Jean-Luc Picard (*Star Trek*), and Martin Riggs (*Lethal Weapon*) are presented as heroes precisely for their ability to free themselves from the identity they mark. Here, they craft a slippery whiteness that is both nonwhite and too white, one that balances this newly marked whiteness with the whiteness of transparency.

As Dyer compellingly tells it, the history of whiteness has been a striving after a perfected, luminescent glow of purity by, among other things, dying the skin with "ceruse (white lead) which made the wearer look matt white and poisoned the skin," a technique used from ancient Greece through the beginning of the nineteenth century (p. 48). Dyer sees this quest towards "luminescence" in the context of "the idea of whiteness as transcendence, dissolution into pure spirit and no-thing-ness" (p. 80). This "no-thing-ness" he links to the ideal of whiteness as self-abstraction: "being nothing at all may readily be felt as being nothing in particular, the representative human, the subject without properties" (p. 80). The new men of this chapter seem to disrupt this no-thing-ness, destabilizing the abstraction of whiteness by making race an explicit topic. In doing so, however, they offer up a whiteness that subsumes the nonwhite others through which they unwhiten their own masculinity, reiterating the ability of whiteness to be everything (and thus no-thing). Here, this unblushing male's new-found "color" serves only to refigure his noncolored, abstract, un-blushing self.

White Amnesia and Apologia

Dances with Wolves ushered in the '90s by suggesting a way in which white characters could take up the plights of "nonwhite others," all to the cheers and celebrations of the U.S. moviegoing public. In the film, Kevin Costner's character, John Dunbar, goes through a transformation of identity that seemingly takes him from dedicated cavalryman, dutifully holding his post in defense of U.S. nationalism, to full-fledged Native American warrior, battling the white man on behalf of the Sioux. Dunbar's transformation is instructive in its attempts to both recognize and reconcile histories of white oppression. The opening scenes of the film frame Dunbar as a model of U.S. nationalism. Sent to an isolated, abandoned cavalry post by an insane commanding officer, Dunbar nonetheless buys supplies, builds camp, and holds the fort like any dedicated U.S. soldier. Though the camp's occupants have clearly been killed, scared away, or sent elsewhere, Dunbar has faith that his fellow soldiers may one day return and devotedly maintains his post as he awaits them. Faithful to his fellow soldiers, as well as to the nation, Dunbar puts duty above his own safety and comfort, sacrificing himself for the good of the United States.

Dunbar's mobile identity, however, allows him to develop a sense of identification with a nearby group of Native Americans, trading, it seems, his faith in U.S. nationalism for a new devotion to the Sioux Nation. Like a curious child hoping to make new friends, Dunbar gradually enamors himself to the Sioux men who visit his fort. Eventually, he welcomes them into his living space, treating them to coffee and sugar. Once he has begun to befriend them in this manner, he decides to visit their space, suiting up in full dress uniform and riding out for a visit. Following these early, awkward meetings, which have Dunbar and his new Sioux acquaintances struggling to feel comfortable in each other's presence, Dunbar is welcomed as a friend and the film begins its focus on the hilarity of bridging the cultural and language gaps that separate them. The Sioux people's comical mispronunciation of Dunbar's name as "Dumb Bear," for instance, serves as evidence of this separation, as do several potentially dangerous exchanges in which Dunbar's misunderstanding of Sioux custom nearly get him into fights (as when he tries to reclaim his hat from a Sioux male who found it on the battlefield).

Of course, none of these cultural differences or problems are too much for Dunbar to overcome and he is eventually granted full status in the Sioux family. From hunting buffalo to warring with enemy tribes, Dunbar participates in each of these Native American people's important rituals. After he kills a buffalo with his rifle, he reenacts the story over and over for the hosts of young Sioux children in awe of his abilities. Likewise, a group of Native children paint Dunbar's horse in preparation for a fight with a rival tribe, allowing Dunbar to fully participate in the rituals of battle. Perhaps nowhere is Dunbar's presumed assimilation into the Sioux lifestyle more clear, however, than when, dressed in full native apparel, he renounces a group of U.S. cavalrymen in his new Sioux tongue. Later, Dunbar kills a white cavalryman who attempts to harm some native people, making explicit his allegiance to the Natives in the face of U.S. nationalism. And when a group of Sioux warriors battle these cavalrymen in order to free Dunbar, they seem to demonstrate that the feelings of allegiance are mutual.

Throughout, however, Dunbar's faith in the United States is never fully shaken. As the film ends, he and his new wife leave their Sioux family, hoping to tell the truth about these Native peoples to "those who would listen." Faithful that such people exist, Dunbar maintains the "goodness" of at least some portion of white America, presumably that segment of which Dunbar himself is part. Indeed, the film seems to suggest throughout, just as there are "good" and "evil" Native peoples (i.e., the Sioux and the Pawnee), so there are "good" and "evil" whites, the good represented by "those who would listen" and the evil represented by those dastardly cavalrymen who would willfully injure an innocent Sioux. Of course, Dunbar's own goodness is of the highest level, as the end of the film makes clear. Willing to leave his Sioux family, despite his desire to stay among them, Dunbar returns to the white world, his ability to pass between whiteness and native-ness apparently the key asset to saving his people (that is, the Sioux).

Dunbar's ability to pass between whiteness and nonwhiteness seems a significant characteristic during a moment of cultural marking, a characteristic he shares with President Clinton as well as with the masculine images discussed here. While Dunbar, Clinton, and these other images certainly did not invent this racial mobility, it achieved a particularly important cachet amid this discourse of masculine crisis. This section explores two interesting moments of this racial mobility, both taken from the post–*Dances with Wolves*, Clinton era of white masculinity, a period marked by the increasingly conflicted racial negotiations discussed above. Both *Man on the Moon* (1999) and *Star Trek: First Contact* (1996), offer bizarre interventions in this history of white oppression, offering up heroes who both highlight and minimize the problems of whiteness. Andy Kaufman (as portrayed by Jim Carrey in *Man on the Moon*) and Jean-Luc Picard (*Star Trek: First Contact*) are made heroic precisely for the instability of their identities. Whether bleaching white Andy Kaufman's Jewishness, so as to offer a new white hero that denies racial difference, or offering a white victim (in Picard) who learns to overcome his victimization, these films use a mobile racial identity to bleach clean the racial subjects with which they deal. In so doing, they return whiteness to the unblushing American male, forgetting and/or smoothing over the histories they seem to foreground.

The Chaotic Lives of Andy Kaufman

Man on the Moon tells the story of Andy Kaufman, seen variously as a comic genius, a bizarre performance artist, and an unbalanced personality teetering on the edge of insanity. Whatever his brand of performance, the Kaufman memorialized in *Man on the Moon* is one who flows in and out of these performative moments from one instance to the next. This Kaufman stages a fight on the *David Letterman Show* at one moment and serves milk and cookies to an audience at Carnegie Hall at another, all the while maintaining a day job as a waiter in a local coffee shop. Scott Alexander and Larry Karaszewski, who wrote the film's script, describe their interest in Kaufman's bizarrely neurotic antics, saying: "We loved the way he bent reality, confusing put-on with fact, and his good and bad sides made for a rich character" (1999, p. vii). As they tell it, "His act was always about challenging perception, mixing mainstream entertainment with performance art" (p. viii).

For Alexander and Karaszewski, Andy Kaufman's chaotic performances translate to a chaotic identity, the "character" of Andy Kaufman as multifaceted and convoluted as his act. In preparation for the *Man on the Moon* script, these writers had interviewed Kaufman's friends, family, and business associates, trying to establish some sense of Kaufman's character. As they explain it, however, this proved difficult:

> Andy had clearly manipulated his life to create different, discontinuous pieces of himself. Thus, he was an introvert and an extrovert. Some described his all-vegetarian diet. Others related his daily breakfast of bacon and eggs. One girl was led to believe that he was a practicing Orthodox Jew. Marilu Henner found him unwashed and panhandling on a New York street corner. (pp. ix-x)

A 1985 article in the *Los Angeles Times*, written a year or so after Kaufman's death from cancer, reiterates Kaufman's fragmented and chaotic character.

> He was so extraordinarily skilled at keeping people guessing that, a year after his death, he had Hollywood out to see if he had pulled the ultimate stunt—the successful faking of his own death. The evening was billed as the "American Cancer Society (Van Nuys Chapter) in association with The Andy Kaufman Memorial Fund ... invite you to attend 'TONY CLIFTON LIVE' (And Guests)." Top price was $100. Everyone, or almost everyone, in

the audience knew that Tony Clifton, the aggressively untalented Vegas lounge singer, had been Kaufman's alter ego, a role played out with such impeccable dead-pan logic that Kaufman never whipped off the wig and glasses to give it away. That was Kaufman's game—to go you one further, to stare down an audience with a deliberately bad act and thereby challenge the rules of the spectator-performer relationship. In theater parlance it's called "demystification," and Kaufman was a master at it. (Christon, 1985, p. 7)

Simultaneously Tony Clifton, Latka on the television series *Taxi*, a vegetarian, a meat eater, an Orthodox Jew, alive and dead, Kaufman's life seems to offer an extreme version of the mobility and no-thing-ness that Dyer discusses above, the no-thing-ness characteristic of a transcendent whiteness.

This no-thing-ness is an important component of *Man on the Moon*, ultimately centering the film's portrayal of Andy Kaufman. His identity split between a host of competing characteristics, Kaufman here appears the bastion of a newly unmarked whiteness, his mobile identity allowing him to flow between racial markings without himself being marked. This chaotic mobile identity is evidenced from the film's beginning, a comical introduction in which Andy Kaufman (Jim Carrey) addresses the audience directly. Speaking in an odd, unidentifiable "foreign" accent, Kaufman urges the audience not to see the film.

> Hallo. I am Andy. Welcoom to my movie. (beat; he gets upset) It was very good of you to come ... but now you should leave. Because this movie ees terrible! It is all LIES! Tings are out of order ... people are mixed up ... what a MESS! (he composes himself) So, I broke into Universal and cut out all de baloney. Now it's much shorter. (beat) In fact—this is the end of the movie. So tanks for comink! Bye-bye. (Alexander and Karaszewski, 1999, p. 1)

With this final goodbye, Kaufman puts on a record and rolls the film's final credits. Once the credits have finished, however, Kaufman loses his mock accent and begins addressing the audience as the "real" Andy Kaufman. "Okay, good," he begins. "Just my friends are left. I wanted to get rid of those other people ... they would've laughed in the wrong places" (Alexander and Karaszewski, 1999, p. 1). With this introduction, Kaufman picks up a Super 8 movie projector and "begins the real film." This opening scene demonstrates the theme of chaotic identity that is carried throughout *Man on the Moon*.

First, by reprising this Kaufman character, known simply as "Foreign Man," the film points towards an ambiguous and unidentifiable "otherness" important to the film's presentation. Likewise, the pretense of "Kaufman's" dissatisfaction with "his" film, demonstrates the sort of "put on" for which Kaufman has become famous, a feature central to the film's memorial. Here, everything is a "put on," from the film itself to the identities through which Kaufman performs, identities as easily put on as shed.

This theme also plays out in the film's depictions of Kaufman's childhood, which present him as an eccentric child out of touch with the realities of both the world and himself. In one scene, the young Andy is "putting on a show" for an imaginary audience in his bedroom. Working to establish Kaufman's love of performance, the scene frames this young Kaufman as lost in his world of imaginary entertainment. As Alexander and Karaszewski's script details it:

> Stanley [Andy's father] hurries up to Andy's shut door. We hear little Andy doing voices.
>
> >ANDY (o.s.)
> >   (as WORRIED GIRL)
> >But Professor, why are the monsters growing so big?
> >   (now as BRITISH PROFESSOR)
> >It's something in the jungle water. I need to crack the secret code.
>
> Stanley rolls his eyes. He opens the door ... revealing ANDY, 8, performing for the wall. Andy is happy and enthusiastic ... as long as he's acting. (pp. 2–3)

Later in the same scene, Stanley chastises little Andy, telling him, "You should be outside playing sports," to which Andy replies "But I've got a sports show. Championship wrestling, at five" (p. 3). After young Andy motions towards his "audience," Stanley rebukes him again, demanding emphatically: "That is not an audience! That is plaster! An audience is people made of flesh" (p. 3). Only enthusiastic "as long as he's acting," and able to confuse the wall for an audience, young Andy is here positioned with the same mobile, chaotic identity for which the film remembers Kaufman's character.

From these opening scenes of Kaufman's childhood in Great Neck, Long Island, the film quickly turns to Kaufman's life as an adult performer for comedy clubs and nightclubs in New York. Here, we see the early versions of Kaufman's Foreign Man (who will later become Kaufman's character, Latka, on the television series *Taxi*), as well as the adult versions of the childhood games he played before his audience of the wall. As is made clear through the club audiences' scattered applause, murmurs, and awkward looks, Kaufman's performances leave the crowds confused, his deadpan "humor" and childhood sing-a-longs leaving them wondering if Kaufman's act is for real—if this is really supposed to be entertainment. As *Man on the Moon* tells it, it is only Kaufman's Elvis impersonation that wins over these early New York audiences, ultimately establishing him for the performance artist that he is. In one scene, Kaufman seems to magically transform from the awkward, frightened Foreign Man to a confident, "dead-on" (to quote Alexander and Karaszewski's scene notes) impression of Elvis Presley. Here, Kaufman rips off a stripe from the side of his pants to reveal a row of rhinestones, dawns a white, jeweled jacket, brushes his hair and strikes an Elvis swagger and pose. Spinning around with a guitar in hand, he curls his lip and begins singing "Blue Suede Shoes." "Flabbergasted," Alexander and Karaszewski explain, "people applaud. This man *is* Elvis" (p. 10).

It is important that the film frames this moment as the turning point in Kaufman's career as a performer, establishing his Elvis impersonation as his first masterpiece of performance art, and thus his big break towards getting himself discovered. Indeed, Elvis Presley is himself tied up with a set of interesting identity and racial problematics that layer onto and complicate Kaufman's own mobile and chaotic identity. The subject of countless contemporary commentaries, biographies, movies, and so forth, the contemporary Elvis persona consists of a series of contradictory fragments and manifestations: heart throb, all-American boy, sexual rebel, overweight Vegas lounge lizard. Hopkins (1971) presents an image of a squeaky clean all-American Elvis, drawing a picture of a Tupelo boy made good.

Elvis was back in Tupelo. It was a model homecoming, so clichéd as to defy credibility. Already Elvis represented the all-American prototype, the Horatio Alger hero of the South: the son of a dirt-poor sharecropper who sang in his mama's church and went on to massive wealth and fame. (p. 160)

In contrast, Goldman (1981)[4] describes an Elvis that is anything but "all American."

> [Elvis] sure as hell wasn't the All-American Boy. Elvis looked nothing like the stock young movie star of the day…. Elvis was the flip side of this clean-cut conventional male image. His fish-belly white complexion, so different from the "healthy tan" of the beach boys; his brooding Latin eyes, heavily shaded with mascara; the broad fleshly contours of his face, with the Greek nose and the thick, twisted lips; the long greasy hair, thrown forward into his face by his jerking motions. (p. 191)

The all-American male by some accounts, and his antithesis by others, Elvis was and is a bundle of important contradictions. A polite young Southern man who loved his mother and said "Thank you," "Sir," and "Ma'am," and a sexual provocateur who shook his pelvis to the screams of young women, Elvis' persona is conflicted in ways similar to Clinton, Kaufman, and the other men discussed in this chapter.

Not the least important of these contradictions is Elvis' negotiation of racial stereotypes and expectations. Recording pioneer Sam Phillips' famous search for a white performer with a black sound ends in the mythic figure of Elvis Presley, a crosser-over of musical and racial boundaries. Whether celebrating his ability to help put race music on the mainstream, white charts, or decrying his theft of African American music, popular discourses on Elvis highlight his problematic racial negotiations in illuminating ways. If Elvis is seen to be "impersonating" an African American style (whether this impersonation is imagined as positive or negative), then "impersonating Elvis" takes on added layers of contradiction and negotiation. *Man on the Moon*'s evocation of Kaufman's Elvis impersonation layers these racial tensions with the contradictions of Kaufman's own chaotic identity. Discussing Kaufman's Elvis, Greil Marcus (1991) highlights these contradictory layers:

> The country is well aware of Elvis as he's been worshipfully caricatured by thousands of Elvis imitators, like the TV comedian Andy Kaufman—who, twice removed, a knowing parody of a parody that doesn't know it is a parody, in some aspect of his act wants to get it across that in fact he loves Elvis, would in some part of his soul give up a limb to feel as Elvis must have felt when he sang as, in his best moments, he sang, and who, as a comedian on TV, in this context of camp layered over bad taste, cannot begin to get such a thing across. (p. 33)

These layers are important as they both evoke and bury the racial background underwriting their performance, "Presley's not-quite and yet not-white absorption of black style" that is "inevitably indebted to a musical tradition of racial impersonation" (Lott, 1997, p. 203).

In evoking Kaufman's Elvis impersonation as the key to his success in being discovered, *Man on the Moon* celebrates Kaufman's ability at "putting on" one of the great "put-er on-ers." For by memorializing Kaufman's Elvis impersonation, *Man on the Moon* references and memorializes a complex chain of reiterated "blackface" performances as well. As Eric Lott explains, "Himself an alleged impostor, the historical referent of Norman Mailer's 1957 'The White Negro,' Elvis inherited a blackface tradition that lives a disguised, vestigial life in his imitators" (1997, p. 203). Kaufman's impersonation of Elvis performs a "double mimicry" of the blackface aura implicit in Elvis' own performance. "Mimicking him" Kaufman, "[impersonates] the impersonator, a repetition that nearly buries this racial history even as it suggests a preoccupation with precisely the blackface aura of Elvis" (Lott, 1997, p. 203). *Man on the Moon* puts on Kaufman putting on Elvis putting on blackface, burying this racial history below multiple layers of parody and repetition, multiply repressing the very history that allows this performance to take place.

This multiple mimicry of blackface is still more important when considered alongside another racial referent all but absent from *Man on the Moon*: Andy Kaufman's Jewishness. While the film's final script describes Kaufman's boyhood home as "an upper-class Jewish neighborhood," this textual reference is, of course, absent from the film itself, as is any other mention of Kaufman's Jewish identity. Although Alexander and Karaszewski's published scene notes contain at least one scene that explicitly references Kaufman's Jewish background—a scene set during Passover in which a young Andy dawns a black beard, hat, and robe, in order to impersonate Elijah—this scene was cut before the final draft of the script. As they explain:

> In our final script, there's only one childhood scene. However, we originally had ten pages of this stuff. From our interviews with Andy's family, we were inundated with interesting, magical stories. It was clearly a way to round out Andy's character: magic, Judaism, eccentric childhood, love of family…. But ultimately, we had to conserve our running time for the main character. He's the one the audience is identifying with. (p. 155)

While the scene may indeed have been cut in order to offer more screen time to adult Andy, this cutting also effectively eliminates reference to Andy's Jewishness.

Taken together, Kaufman's parodic "blackface" and absent Jewishness bring *Man on the Moon*'s implicit theme of whiteness into dramatic relief. From his impersonation of Elvis' impersonation of African American performance to his ambiguous "Foreign Man" character (from Caspiar, a "veddy small island in de Caspian Sea" [Alexander and Karaszewski, 1999, p. 11]), the Kaufman of *Man on the Moon* constantly puts on the face of some Other. In his study of the history of blackface minstrelsy and Jewish identity, Michael Rogin (1996) explores ways in which blackface performances have served historically to "Americanize," and, indeed, "whiten" Jewish immigrants, an idea that sheds light on the racial dynamics of *Man on the Moon*. Discussing *The Jazz Singer*, for instance, Rogin argues that "With Jolson cheering for 'my mammy' and Uncle Sam, blackface as American national culture Americanized the son of the immigrant Jew" (p. 6). Elaborating this process more fully, Rogin asserts that:

> In the hands, disproportionately, first of Irish and then of Jewish entertainers, this ethnocultural expression served a melting-pot function. Far from breaking down the distinction between race and ethnicity, however, blackface only reinforced it. Minstrelsy accepted ethnic difference by insisting on racial division. It passed immigrants into Americans by differentiating them from the black Americans through whom they spoke, who were not permitted to speak for themselves. Facing nativist pressure that would assign them to the dark side of the racial divide, immigrants Americanized themselves by crossing and recrossing the race line. (p. 56)

Because "whites who black up call attention to the gap between role and ascribed identity by playing what, in the essentialist view, they cannot be" (p. 34), blackface has historically served to whiten immigrants, placing them against the black Americans whom they presumably mimic.

A similar whitening up takes place with *Man on the Moon*'s presentation of Andy Kaufman, though it is a whitening up of a different cultural moment. In one particularly important scene, one that Alexander and Karaszewski describe as the "secret to the movie" (p. xii), Kaufman briefly laments to his girlfriend, Lynne, the public's misunderstanding of and hostility towards his personality. "I like you," Lynne replies, trying to comfort him. "You don't know the real me,"

Andy responds. To which Lynne quickly observes: "Andy ... there *is* no real you" (p. 118). Of course, Kaufman agrees, as do Alexander and Karaszewski. Always in Other faces, always performing some Other identity, there is no real Andy Kaufman as far as *Man on the Moon* is concerned. A practitioner of Transcendental Meditation, a staunch vegetarian, a meat eater, lounge singer Tony Clifton, a Christian,[5] Andy Kaufman is here but a collection of random identities strung together haphazardly. Like the Jewish performer in blackface and similar to Kevin Costner's taking up of Native American identity, *Man on the Moon* offers up a masculinity in Other face, though one made more radically chaotic through the presumed chaos of Kaufman. Kaufman dawns Other face after Other face, consistently discarding each of the identities he takes up and leaving "himself" an empty, identity-less hole in their absence.

Alexander and Karaszewski describe their film as an "antibiopic—a movie about someone who doesn't deserve one" (p. vii) and, indeed, as framed by *Man on the Moon*, Kaufman is a nothing. This is the very nature of *Man on the Moon*'s remembrance of Kaufman, the very basis for its heroic retelling of his antiheroic life. Split between multiple identities, Kaufman is himself magically freed of the burden of identity, offering up a new, universally generalizable subject. In the end, all that's left is the blank no-thing-ness of white.

Turning the Tables on History in Star Trek

The media phenomenon of *Star Trek*, in all of its myriad forms, has received much recent scholarly attention. Henry Jenkins' book *Textual Poachers* (1992) offers an interesting celebration of *Star Trek* fandom, exploring the ways in which *Star Trek* fans "poach" from the program, using fragments of the show to create aberrant readings that often run against its dominant ideological themes. In the wake of Jenkins' book, others have worked to explore the politics of identity played out in *Star Trek* as well, interrogating the multiple identities at play in *Star Trek*'s multiple programs (see, for instance, Gregory, 2000; Bernardi, 1998; and Harrison et al., 1996). This next section continues this exploration with a focus on the two trajectories of identity important to this chapter, whiteness and, of course, masculinity, as played out in a recent evocation of *Star Trek: The Next Generation* film, *Star*

Trek: First Contact. Similarly to *Man on the Moon*, I argue, this film re-asserts the invisibility of whiteness, here by "victimizing" and then "rescuing" the body of their white male lead, Jean-Luc Picard.

These conflicted politics of identity have continuously been central to *Star Trek*, its multicultural, multiethnic, multi-life-form messages playing an important role since the show's inception. The multicultural crew of the original *Star Trek*'s U.S.S. *Enterprise* evidence a utopian dream of racial and cultural harmony, the future fantasized by the show's creator, Gene Roddenberry. The competing peoples and governments of Earth replaced by a Federation of united countries, states, and nations, the *Enterprise*'s crew is a vignette of this fantastically united Earth. Staffed by an African American woman, Lieutenant Uhura; an Asian man, Mr. Sulu; and a Russian man, Ensign Chekov, the *Enterprise* promises a post–Cold War, post–Civil Rights future of peace and togetherness, though all under the charge of the white, Midwestern Captain James T. Kirk. The offspring of the Federation and bent on seeking out "new lives and new civilizations," the U.S.S. *Enterprise* is a utopian mission of ethnographic curiosity, the outstretched hand of a united Earth hoping to unite with others.

Of course, as multicultural as it attempts to be, this early *Star Trek* cannot rid itself of its paranoid fear of the very otherness it presumes to embrace. While it offers a utopian vision of racial and cultural unity, it simultaneously undercuts this vision in often far from subtle ways. A telling incident is the show's final episode, "Turnabout Intruder," aired June 3, 1969, in which Captain Kirk is forced to switch bodies with Janice Lester, an apparently overly ambitious woman who is envious of Kirk's power. Using an ancient alien device, Dr. Lester trades bodies with Kirk, then assumes command of the *Enterprise*, leaving Lester's body (now Kirk) trapped in the ship's infirmary. As the "real" Captain Kirk struggles to escape and to explain Lester's diabolical actions to the crew, Lester makes plans to kill Kirk and assume his identity forever. Before he can be killed, Kirk manages to escape the infirmary and explain the bizarre feat of body swapping to which he has been subjected. Dubious about this strange tale, the crew and Lester nonetheless agree to hold a tribunal in which Kirk can tell his story. Kirk's "testimony" serves not only to uncover "the facts of the case" but also to condemn Lester's thirst for power. When asked why Lester would perpetrate such a crime, Kirk (in the body of Lester) explains: "To get the power she craved. To attain a position she doesn't merit by temperament or training. And most of all, she wanted to murder James Kirk, a man who once loved her, but her intense hatred of her own womanhood made life with her unbearable."

Here, this early *Star Trek* reacts hegemonically against the changing, "liberal" social values of its moment—the same values outwardly embraced in its multicast crew. Dr. Lester's desire to kill Kirk and take over his role offers a caricature and condemnation of the feminism of the late '60s, evoking a fear of powerful, power hungry women. Captain Kirk's testimony clearly decries Dr. Lester's thirst for power (no doubt further evidence of her intense hatred of her own womanhood) and her desire for the position she does not and, the program suggests, ultimately should not have. For in the end, Dr. Lester cannot replace Kirk, evidence, it seems, of the misdirection of this movement to empower women. Dr. Lester's inability to handle the stress of command eventually overcoming her, she collapses on the bridge, reversing the bodily transference she had forced upon Kirk. The message seems clear: women want to kill men and take their jobs, but ultimately they can't handle them. Returned to his proper body, Captain Kirk is free, once again, to command the ship in an appropriately masculine manner.

Star Trek: The Next Generation (*TNG*) and the film *First Contact* share this uneasy negotiation of "liberal" and "multicultural" values, though theirs is a negotiation of a different historical epoch. While the first *Star Trek* may have blatantly rejected the possibility of a powerful female leader, maintaining the hypermasculine authority of Captain Kirk throughout, in, and after *TNG*, this masculinity is less explicitly and unequivocally asserted. Still more multicultural and inclusive, the *TNG* crew includes more women, as well as a host of "alien" life-forms, building a diversity of the intergalactic variety. The duties of ship medical officer, for instance, carried out by Dr. Leonard "Bones" McCoy on the original *Enterprise* are taken over by Dr. Beverly Crusher for *TNG*, and the crew includes Worf, a Klingon, the original program's archenemies, as well as Data, an android lieutenant commander. *Star Trek: Deep Space Nine* and *Star Trek: Voyager*, both offspring of *TNG* and contemporaries of the film *First Contact*, carry these multicultural themes still further, presenting an African American man (*Deep Space Nine*) and a white woman (*Voyager*) in command of the ship. If the original *Star Trek* worked at multicultural themes, these new programs seem to have made these a central and driving concern.

Still, just as the original *Star Trek* of "Turnabout Intruder" could not free itself from the dominant masculinity of its era, so these newly enlightened treks negotiate the conflicted masculinity of the '90s in conflicted and troubling ways. In the same way that the original *Star Trek* plays out a masculine angst on the body of James T. Kirk, so *TNG* film *First Contact*, a product of this new, seemingly more enlightened moment, works out a similar anxiety on the body of its white male lead, Jean-Luc Picard. If "Turnabout Intruder" is primarily about the infiltration of feminism and powerful women into the public sphere, however, then *First Contact* concerns itself with the penetration and defilement of traditional masculinity itself. In "Turnabout Intruder" the question of white masculinity is itself never broached. Even in the body of a woman Captain Kirk is his rational, calm, deliberate self—evidence, it seems, of the essentialized soul of maleness that Dr. Lester, and, indeed, any woman, can never achieve.

In *First Contact* Picard's vulnerability and the fragility of his self forge an explicit theme, the movie centered on Picard's journey to reconcile himself with a bodily violation perpetrated against him by an alien species—an assimilating mass of technological bodies known as the Borg. Part machine and part organic, the Borg search the universe for new life-forms to assimilate into their collective, turning autonomous creatures into mere extensions of their cybernetic mass body. Picard has a history with these Borg. Having been assimilated into their collective in the past, he still bears the physical and mental scars of this bodily occupation. Continuing this story line, *First Contact* explores Picard's relationship to the Borg, presenting his attempt to reconcile with these horrible memories and to reestablish the autonomy of his own body. In so doing, I argue below, *First Contact* also reestablishes a sense of universal, unblushingly white masculinity. Whereas "Turnabout Intruder" recenters a universal masculinity through an explicit condemnation of women "doing man's work," *First Contact* does so by penetrating, and, indeed, violating the body of its white male commander. By victimizing its white male lead, at the same time that it celebrates and reiterates a history of liberal, multicultural values, *First Contact* works to salve the racial history that would challenge whiteness and traditional masculinity in the first place, rereading this history in a way that recaptures a sense of the no-thing-ness of whiteness.

First Contact introduces Picard's struggle with the Borg and his fear of bodily penetration from its opening scenes. These opening shots make up a flashback sequence in which Picard remembers his traumatic experience of being imprisoned on the Borg ship—a flashback that moves from Picard's initial capture to his full-fledged assimilation into the Borg collective. Beginning from an extreme close-up of Picard's eye, as if to suggest a view from deep within his psyche, the camera pulls back to a full body shot, showing Picard strapped upright into the mechanical contraption in which he is imprisoned. From here, the camera pulls back still further, revealing the enormity of the Borg space craft, a gigantic technological beast of which Picard is but a tiny piece. Accompanied by the electronic hums of the soundtrack, the scene cuts to a hallway of the ship, the Borg crew moving around rapidly and mechanically. "Human bodies" fitted with vacuum tubes and metallic, computerized pieces, the Borg are a frightening hybrid of computer and person. Depicting Picard's own traumatic assimilation into this cybernetic life, the camera highlights his agony and pain, focusing on his wincing expression in one scene, and offering a close-up of a drill bit penetrating his pupil in another. Finally, the camera reveals Picard in full "Borg dress," his body a suit of black metallic armor and his face half covered in titanium. "I am Locutus of Borg," he says, facing the camera. "Resistance is futile."

This opening sequence frames the movie, graphically representing Picard's traumatic assimilation by the Borg and the powerful ways in which it has implanted itself within his memory. Representing this sense of remembrance, the scene cuts from this shot of "Picard the Borg" aboard the Borg ship, to a shot of Picard in his Star Fleet uniform, aboard the *Enterprise*, presumably thinking back on his experiences of assimilation. Staring into space, Picard seems lost in these memories. He sighs, runs water over his hands, then splashes his face, presumably attempting to relax after his traumatic dream. As he looks into the mirror, however, a silver device explodes from his face, its metallic extensions implanting themselves on his cheek. As the camera cuts to Picard suddenly sitting up in bed, this mirror sequence, like the flashback that precedes it, is revealed as a dream. A dream about remembering his time with the Borg, this sequence emphasizes still further its focus on Picard's recollection of his past trauma. How Picard will negotiate his personal history with the Borg is one of the primary questions of the film, a question posed dramatically from this opening sequence.

Another question of history concerns a second, connected story line in *First Contact*, also related to the Borg. Bent on assimilating the entirety of the human race into their collective, the Borg send a series of ships to Earth. While Picard and the Enterprise are perhaps best equipped to battle them, the Star Fleet commanders order them to a neutral part of the galaxy. As Picard tells it, "They believe that a man who was once captured and assimilated by the Borg should not be put in a situation where he would face them again. To do so would introduce an unstable element to a critical situation." However, when the Star Fleet crews are unable to deflect the Borg attack, Picard violates his orders, taking the *Enterprise* and its crew to participate in the battle. Once there, they defeat the Borg attack, Picard's knowledge of the Borg helping them destroy their main ship. Before they are destroyed, however, the Borg send a ship back in time, planning to destroy Earth in the past, before the Federation has a chance to develop and unite humanity across the globe. Without this unity, the Borg know, Earth will be an easy target for conquest. By destroying Earth before the moment of "First Contact," when Earthlings first make contact with an extraterrestrial life-form, the Borg will prevent the historic moment that gave rise to the utopian world of multiculturalism and global unity that informs the *Star Trek* universe. Bravely, Picard and his crew follow the Borg back into the past, destroying the Borg, but not before they can do significant damage to the Montana site of this first contact—damaging the spaceship that will fly into outer space, catching the attention of the aliens who first visit Earth. The *Enterprise*'s crew must help repair the ship, ensuring that its historic flight takes place and rescuing their future history of utopian togetherness.

This story line allows *Star Trek* to retell its own utopian history, stitching together the logic that leads to its idyllic world of equality. The return to 21st-century Montana is a return to a pre-united, pre–*Star Trek* Earth, an apocalyptic vision of life after World War III. This is the site where Dr. Zephram Cochrane, a scientist experimenting with travel beyond the speed of light, fashions a spaceship out of an old nuclear missile and scraps of discarded titanium. Shabbily dressed and drunken, Dr. Cochrane seems uncharacteristically ordinary for someone who will be memorialized and celebrated as much as three centuries in the future. Indeed, Cochrane and his 21st-century friends are astonished by the awe with which these 24th-century time travelers revere them. "I went to Zephram Cochrane High School," one of the *Enterprise*'s crew tells Cochrane, while another stammers nervously as he musters the courage to shake Cochrane's hand. After being told that he is standing on the very spot where his statue will be erected, Cochrane runs away, his ordinary personality unable to handle this extraordinary praise. After much persuasion Cochrane reluctantly agrees to help repair his ship and fly this historic mission. As Deanna Troi, one of the *Enterprise*'s crew explains it, "It unites humanity in a way that no one ever thought possible when they realize they're not alone in the universe. Poverty, disease, war, they'll all be gone in the next 50 years." With Zephram Cochrane, *First Contact* creates a white American antihero to accomplish the most extraordinary of feats, unheroically and unintentionally uniting all of Earth's occupants.

This accidental, unintentional unification stands in stark contrast to the practices of the Borg, who intentionally assimilate others, like Picard, against their will. While crewmembers on Earth convince Cochrane to fulfill his accidental destiny, Picard remains on board the *Enterprise* with the remaining crew, battling Borg who infiltrate the ship as it travels back to the 21st century. This battle both demonstrates the ruthless assimilative practices of the Borg and offers Picard a chance to negotiate his own painful personal history with these ruthless colonizers. Preparing his crew to fight the Borg on board the *Enterprise*, Picard instructs them that they "may encounter *Enterprise* crewmembers who have already been assimilated," adding, "don't hesitate to fire, believe me, you'll be doing them a favor." Still dealing with his own painful memories, Picard is caught up in his own trauma and resentment towards the Borg. In one scene, Picard kills a Borg, screaming as he shoots him repeatedly with his gun. Looking at the body, Lily, a 21st-century African American woman brought aboard the *Enterprise* to recover from radiation poisoning, notices the Star Fleet uniform on the dead Borg body. "It's one of your uniforms," she tells Picard sadly. "Yes," Picard replies coldly. "This was Ensign Lynch." "Tough luck, huh?" Lily replies sarcastically, surprised by the lack of compunction Picard demonstrates for this former crewmember. Angry about his past experience with the Borg, a dead crewmember is a small price to pay for destroying part of the Borg collective.

Picard's profound trauma in facing the Borg is most evident in a confrontational scene between Lily and Picard in which Lily warns Picard of the dangers of harboring such resentments. As the Borg continue to assimilate the Star Fleet crew, the remaining crewmembers urge Captain Picard to abandon the ship and destroy it with the remaining Borg inside. "No!" Picard responds vehemently. "We have to stay and fight. We are not going to lose the *Enterprise* ... not to the Borg ... not while I'm in charge." While the crew reluctantly takes Picard's orders as the final word, Lily confronts Picard and challenges him to consider his personal history with and investment in this battle with the Borg. Says Lily:

> Okay. I don't know jack about the 24th century. But everybody out there thinks that staying here and fighting the Borg is suicide.

"None of them understand the Borg as I do," Picard responds. "No one does. No one can." "What is that supposed to mean?" Lily challenges. In response, Picard reiterates the painful process of assimilation to which he has been subjected:

> Six years ago they assimilated me into their collective. I had their cybernetic devices implanted throughout my body. I was linked to the hive mind, every trace of individuality erased. I was one of them. You can imagine, my dear, I have a somewhat unique perspective on the Borg, and I know how to fight them.

Rather than buy into Picard's reasoning, however, Lily uses this story to suggest the irrationality and pointlessness of his quest to destroy the Borg. "I am such an idiot," Lily responds, "It's so simple. The Borg hurt you and now you are going to hurt them back." Frustrated with Lily, Picard retorts, "I don't have time for this." "Oh hey," Lily answers back sarcastically again, "I'm sorry, I didn't mean to interrupt your little quest. Captain Ahab has to go hunt his whale." With this, Picard screams in frustration, smashing a display case filled with miniature replicas of Star Fleet spaceships. Enraged, he continues his rationalization:

> I will not sacrifice the *Enterprise*. We've made too many compromises already, too many retreats. They invade our space and we fall back. They assimilate entire worlds and we fall back. Not again. The line must be drawn here! This far, no farther. And I will make them pay for what they've done.

Lily, with a look of concern and pity on her face, picks up part of the smashed display case, one of the replica spaceships it had contained. "You broke your little ships," she says. "See you around, Ahab." As she turns to leave, Captain Picard suddenly goes through an amazing metamorphosis, as if Lily's last comment has suddenly opened his eyes to his own frustrated struggle with the Borg. Staring off into space, Picard begins reciting:

> "And he piled upon the whale's white hump a sum of all the rage and hate felt by his race. If his chest had been a cannon he would have shot his heart upon it." Ahab spent years hunting the white whale that crippled him—a quest for vengeance. But in the end it destroyed him and his ship.

Alongside Lily's persuasive comparison to Captain Ahab, Picard finally sees his anger towards the Borg for what it is, an obsessive, irrational, and dangerous quest for revenge. His eyes opened, he instructs the crew to evacuate the ship and makes plans to destroy it—finally, it seems, at peace with his past suffering at the hands of the Borg.

Together, these fragmented but complementary story lines—the assimilative practices of the Borg, Picard's frustration and realization via his confrontation with Lily, the antiheroics of Zephram Cochrane—all weave an interesting story of white masculinity. The Borg, for instance, with their relentless quest to assimilate others, seem a stereotype of whiteness itself. As Rhonda Wilcox (1996) explains:

> The Borg begin fictional life as a condemnation of white policies of "assimilation" (their term). In the well-known two-part episode "The Best of Both Worlds," they attempt to take over the Federation's various humanities, declaring every individual difference "irrelevant" and killing thousands in a grotesque attempt, as they put it, to "raise the quality of life," a harsh parody of white assimilationist and colonialist practices and an implicit endorsement of "race cognizance" by contrast. (p. 79)

First Contact foregrounds this harsh critique of whiteness still further. In a scene in which the android Data has been captured by the Borg, the "Queen Borg" lays out their assimilative, colonial practices and the "logic" that drives them. When Data asks this Queen Borg who she is, she responds, "I am the Borg," to which Data responds, "That is a contradiction. The Borg have a collective consciousness. There are no individuals." "I am the beginning, the end, the one who is many," the Queen Borg explains, describing explicitly the unity in one that the Borg seem to represent, the harsh caricature of assimilation Wilcox describes above. Continuing, this Queen Borg explains that she "bring[s] order to chaos," describing her role in the Borg's "quest to better ourselves. Evolving towards a state of perfection." Responding, Data challenges that "the Borg do not evolve, they conquer." To which the Borg Queen asserts, "By assimilating other beings into our collective we are bringing them closer to perfection." This obvious disregard for other life-forms on the road to the Borg's "quest for perfection" constitutes *First Contact*'s explicit critique of white assimilationist practices, its seemingly biting caricature of colonialist whiteness.

But just as "Turnabout Intruder" needed to rescue itself from its newly liberalized cultural values, so *First Contact* rescues itself from this presumed critique of whiteness—through both the "historical tale" of Zephram Cochrane and Picard's personal history with the Borg laid out above. First, in telling the tale of Zephram Cochrane, *First Contact* reiterates the utopian ideals at the heart of *Star Trek*'s futuristic fantasy. By rescuing Cochrane's ship, the *Enterprise* crewmembers help guarantee a future of unification and peace. Although the Cochrane/Federation story is, significantly, a tale of white America—of a white Montana man who unites the world, however unintentionally—*First Contact* positions this story in stark contrast to the tale of the ultra-white Borg who assimilate others without any concern for their individuality or will. The Federation unifies, the story goes, while the Borg assimilate. In the same way, Picard's hatred towards, and eventual coming to terms with, the practices of the Borg further this complicated picture of whiteness. Following the criticisms of the African American character Lily, Picard realizes the irrationality of his hatred towards the Borg, finally recognizing the necessity of leaving this animosity in the past. In the process, the American whiteness of the implicit subtext is rescued. If an African American woman can recognize the irrationality of hating the ultra-white Borg, then can't we all, with Picard's help, see the futility of criticizing the good American whiteness, particularly since it unifies rather than assimilates?

These complex juxtapositions work to unwhiten the white, nationalist fantasy that underwrites the *Star Trek* utopia. In the glare of the ultra-white Borg, everything else appears a tame hue of neutral. Here, *First Contact* creates its own version of unblushing masculinity.

The all-English-speaking *Enterprise* and the Federation of planets founded on Montana soil appear spontaneous instances of unity and togetherness. The call to join this unity appears not as a colonialist effort but rather as the inevitable consequence of recognizing our shared characteristics—characteristics not surprisingly white and American. Not unlike the fantastic logic of the World Trade Organization, World Bank, or International Monetary Funds,[6] the fantasy of white America is here seen as the fantasy of the world (indeed, the universe). This is unity, not assimilation, evolution not conquering. Here whiteness is rescued at the same time that it is destroyed, the film's biggest victim, the white male Jean-Luc Picard, able to look past the terrible tortures of his past. Returned to the proper command of the ship, the newly unblushing Picard is able to fly the *Enterprise* into its future of utopian unification and togetherness, free from the awful markings of whiteness and assimilation.

Policing Whiteness

By all accounts, the prime-time drama *Hill Street Blues* revolutionized the vision of the police force offered up by American television. *Dragnet*, for instance, had offered an overly rational, reserved image in police officers Joe Friday and Jack Webb, the clean-cut upstanding policemen who always got their man. *Barney Miller* depicted a madcap, humorous police force of comic relief. *Hill Street Blues*, in contrast, created a messy show of blurred boundaries and blurry characters. Neither fully rational nor madcap, neither comical nor entirely serious, the characters and plots of *Hill Street Blues* depicted a relatively confused vision of police ability and morality. Whereas previous police officers had been either serious, dedicated officers whose ability to solve cases was never in question, or well-meaning comics whose ineptitude was simply part of their hilarious characters, the police officers of *Hill Street Blues* struggle through both their jobs and their daily lives in a variety of powerful ways. As Caren Deming (1991) argues:

> In contrast to the heroes of romantic crime dramas, whose moral and physical superiority are beyond question, the main characters of *Hill Street Blues* are more ironic. They face ambiguous moral problems; and they sometimes weaken in the face of adversity. The struggle for order and sanity occurs,

therefore, within and among them as well as between them and the chaotic world of the streets. Yet, despite apparently overwhelming odds, they never lose their essential goodness. They keep trying, and that makes them heroic. (p. 255)

Their personal and professional lives often derailing in poignant ways, the police officers of *Hill Street Blues* carry on nonetheless.[7]

This ongoing, ordinary life struggle translates into interesting dynamics of masculinity as well. Through his interviews with Steven Bochco and Matthew Kozoll, the show's creators, Todd Gitlin (1983) elucidates the bizarre sort of "masculine heroism" put forward by *Hill Street Blues*:

> "We raised unanswerable questions," Kozoll reflected. "[Network executives] don't like that." In one early show, the recovering Renko, after his gaze at death, hit the skids when the junkie who had shot him was freed for lack of sufficient evidence. Renko began drinking on the job. In uniform, he was seen carrying on with a couple of whores in a restaurant. In Bochco's pungent words, "He was making an asshole of himself." Network executives, he recalled, "got really upset. This is a hero." How could he be permitted to carry on so relentlessly? Then there was the episode in which, shortly after a minor character's death of a heart attack, the fifty-five-ish Sgt. Phil Esterhaus decides to marry a high school girl.... Bochco: "Someone going through a crisis, a midlife crisis, compounded by the death of a fellow officer from natural causes, fifteen years younger than himself—yeah, a guy can act rather foolish." (p. 303)

Creating these sorts of antiheroes, *Hill Street Blues* offers a vision of police masculinity that departs from the squeaky clean, morally upright image of police dramas past. Struggling to maintain their masculinity in the face of a changing world, the *Hill Street Blues* men make do as best as they can, even if that means drinking on the job and marrying high-school-age women.

This is the police world inherited by the '90s television program *The X-Files* and the '90s film *Lethal Weapon IV*. In a world in which cops must struggle not only to do their jobs but to maintain their sense of self, Fox Mulder of *The X-Files*, and Martin Riggs of the *Lethal Weapon* series are in constant states of flux and disorder, both professionally and personally. Indeed, both of these officers constantly have their sanity questioned by those around them. Mulder is branded "Agent Spooky" for his unrelenting belief in the alien abductions and bizarre psychic phenomena assigned to the X-Files moniker. Riggs has been passed off as psychotic for his bizarre and dangerous attempts to catch criminals (in an early film, for instance, he handcuffs himself to a suicidal man and jumps with him off a building). Like the *Hill Street Blues* police officers, however, Mulder and Riggs maintain an essential goodness, framed as heroes of their respective worlds not simply despite, but because of, their bizarre beliefs, actions, and negotiations. Like fellow heroes Andy Kaufman and Jean-Luc Picard, the heroics of Mulder and Riggs depend upon the chaos of their identities—a chaos that allows them a complex negotiation of race, in particular. Linked to some sort of racial other, Mulder to a mysticism *The X-Files* often associates with native cultures, and Riggs to his African American partner Sergeant Murtaugh, these heroes perform a colonialist act of association that seeks to mitigate their chaotic identity via a universal, unblushing whiteness.

Fox Mulder's Vulnerable Body

The contemporary sci-fi drama *The X-Files* offers an interesting inversion of traditional gender stereotypes. Traditionally and stereotypically, rationality has been associated with masculinity (indeed, white masculinity), while such nonrational characteristics as intuition, emotion, and sensitivity have been assigned to a feminine or, as noted in previous chapters, more generally Other category.[8] Thus, the more seemingly rational, analytical, logical domain of the surgeon has stereotypically been coded male, whereas the seemingly sensitive, compassionate world of the nurse is traditionally imagined female. An old riddle depends upon this stereotyped gender coding: A man and his father are injured in a car wreck and are taken away by ambulance. The father dies on the way to the hospital, but the son is taken to the emergency room for surgery. The surgeon, however, upon seeing the younger man, exclaims: "I can't operate. This man is my son!" How can this be? The surgeon, of course, is the man's mother, an obvious answer obfuscated by this gendered logic of profession and domain.

The X-Files seems to invert this logic of gendered domains, challenging these traditional views of masculinity and femininity by casting its two main characters against the grain of these stereotypes. As suggested above, Fox Mulder, the principal male, is framed in communion with an otherworld of psychic and mystical phenomena. Preferring the more bizarre, ethereal answers to the crimes he investigates, Mulder seems an agent of the nonrational. Shown as empathizing with and understanding the complicated world of the psyche, Mulder's explanations for the bizarre crimes he interrogates appreciate the worlds of the occult, the extraterrestrial, spirituality, and superstition, as well as the complex possibilities of the human spirit when put under excessive emotional or psychical pressures. In contrast, his female counterpart Dana Scully is a trained scientist and doctor who prefers the more analytical, rational answers to the crimes the two explore. Hired by the FBI to debunk the bizarre, unorthodox claims of Mulder, Scully's job is to subject Mulder's nonrational methods to the rigors and reasons of science. Always in search of the scientific explanation, Scully gives voice to this stereotypically masculine domain of reason, working to refute Mulder's more emotional, psychical claims. Ultimately, however, Scully must acknowledge, episode after episode, that her scientific methodologies and thinking are not enough to explain the bizarre occurrences assigned to the X-Files. In the end, Mulder's intuition and psychic interpretations prove more reliable than Scully's science.

This gender reversal and its implied chaotic identities make for an interesting and conflicted masculinity in the character of Mulder. As *The X-Files* frames his background, Mulder's sister disappeared when he was a youth, leaving him with an emotional scar he has carried to adulthood. Suspecting that aliens abducted her, Mulder's brotherly love eventually drives him to pursue the X-Files of the FBI, bizarre unexplained cases often written off as unsolvable. On the margins of the FBI, Mulder is an outcast, his chaotic marginalized identity demonstrated throughout his character. Given this marginalized status, it is easy to read Mulder as a critique of traditional gender roles, his association with the nonrational, antiscientific world suggesting this obvious possibility. Joe Bellon (1999) reads *The X-Files* as an act of rebellion against authority, reading Mulder as a concomitant rebellion against traditional masculinity:

> For his part, Mulder no more signifies the traditional male cop than Scully signifies the traditional female. He is prone to strange moods, strong emotions, and light-hearted comments. Unlike more traditional male cop characters who flaunt a devil-may-care attitude as a kind of macho display, however, Mulder represents the emotional and empathic balance to Scully's logic and rationality. He cries over the abduction of his sister when he was a

child ("Conduit"), he empathizes with families who lose their children, and he is obsessed with truth. (p. 150)

Given this, and other rejections of tradition, Bellon claims that "against the monolith of authority, *The X-Files* presents a subversive, liberating vision" (p. 152). While this reading is intriguing, it fails to address the multiple dimensions of identity on *The X-Files*, ignoring in particular the ways in which discourses of race complicate these discourses of gender and authority. A close look at a set of three episodes of *The X-Files* ("Anasazi," "The Blessing Way," and "Paper Clip") demonstrates the complex negotiation of race within Mulder's masculinity, a negotiation that reauthorizes whiteness in peculiar and important ways.

"Anasazi" served as the season finale for the second season of *The X-Files* and "The Blessing Way" and "Paper Clip" were the first and second episodes of season three. Because the shows close one season and open the next, their stories are particularly important within the larger narrative of *The X-Files*, weaving a tale intended to keep fans interested throughout the summer months. As a result, the episodes were not self-contained, but both continued exegesis developed in past episodes and foreshadowed story lines to be continued in the future. Thus, these episodes offer a kind of framework for a larger, overarching *X-Files* narration. Additionally, because of the intriguing racial narrative that informs these three episodes, they tell an interesting story of white masculinity as well, one that resonates with the other stories explored throughout this chapter. This set of episodes opens on a Navajo reservation in New Mexico, setting the stage for a tale of Native American mysticism. These opening scenes take place inside a Native American family's home. As the camera focuses on a young man in bed, an earthquake shakes the family's house, waking the young man up and sending him into the kitchen where two older men are calmly having breakfast. "Eric, leave the snakes alone today," the eldest man, Albert Hosteen, tells the youth as he leaves the house. "They'll be angry and afraid." Then, addressing another man in Navajo (with English subtitles), the elderly man continues, "The Earth has a secret it needs to tell." Riding his motorcycle from the house and into the desert, Eric, the younger Navajo man, comes across "the Earth's secret." Buried in the ground he finds a large metallic object, some enormous container concealed in the ground. He brings the elder Albert and some other Navajo men to see it and together they uncover a strange, deformed skeleton. "It should be returned," Albert says of the skeleton they have removed. "They will be coming." This opening scene frames this three-part *X-Files* narrative, suggesting the complicated tale these three episodes will tell.

At the same time, in Washington, D.C., a computer hacker, codenamed "the Thinker," has broken into a government computer and stolen the Defense Department files that will eventually involve Mulder in this complex ordeal. A conspiracy theorist looking for evidence that the government has known of and concealed information about extraterrestrials, the Thinker has found the "MJ files," computer documents that detail visits from these aliens and the Defense Department's attempts to cover them up. Fearing his life will be in danger, the Thinker arranges a secret meeting with Agent Mulder. Knowing Mulder's history for investigating extraterrestrial and other bizarre phenomena, the Thinker gives Mulder a disk containing these MJ Files, hoping he will use them to expose the government's cover-up once and for all. Mulder is excited, confident he has finally found the hard evidence for which he has been searching for many years. When he opens the file, however, he is disappointed and angry, exploding in an uncharacteristic display of rage. The disk seems to contain gibberish, nonsense strings of consonants and meaningless collections of letters. Agent Scully, however, recognizes the characters as Navajo, apparently a common form of encryption used during World War II. They will need to have the disk decoded to determine if it contains evidence of alien contact with Earth, as well as evidence of a vast government conspiracy to keep this contact a secret.

Even before the disk can be decoded, it becomes obvious that certain members of the government do not want its contents to be discovered. As soon as the Thinker accessed the file, this group of government men had busied itself with the task of recovering this important disk. Dressed in dark suits and depicted in dimly lit rooms, this group of older white men seems the embodiment of "the Man," the stereotypically faceless, nameless, rich, and powerful men responsible for running the state. Enhancing the identity-less-ness of this body of men, the most conspicuous member of this consortium (simply known as Cigarette Smoking Man to Mulder and Scully) is himself *The X-Files'* continuing evocation of this mythically powerful body. Throughout, this nameless body of men makes decisions with little or no regard for the people outside their group. In their attempts to cover up the contents of the MJ Files, they murder witnesses, kidnap innocents, and even mug their own FBI agents. Once they learn that Mulder has received the MJ Files document from the Thinker, they make Mulder a target by publicly charging him with insubordination and dismissing him from his position with the FBI, while privately planning his murder. Like the Borg of *Star Trek: TNG*, this body seems a blatant stereotype of whiteness, an identity-less blob with little concern for those outside of it.

Complicating the narrative of this consortium, we soon learn that Agent Mulder's father, Bill Mulder, himself a former FBI agent, has connections to this group of men. After Mulder receives the MJ File computer disk from the Thinker, the Cigarette Smoking Man visits Bill, hoping that he can deal with his son before he decodes and spreads word of the MJ Files. Bill calls Agent Mulder, asking him to come over to his house and talk about the situation. At his house, Bill explains to his son the delicate nature of the information on the computer disk he has received, warning Mulder that if he investigates the MJ Files and their implications he will expose himself to a set of secrets that many people want hidden. Before Bill can explain these secrets in detail, however, he goes into the bathroom and is shot by a hit man working for the Cigarette Smoking Man. Hearing the shots, Mulder runs into the bathroom and picks his father up in his arms. As he dies, Bill looks into his son's eyes and says, "Forgive me," referring, no doubt, to his past dealings with the consortium. Angered by his father's death, Mulder eventually finds the killer sneaking outside Mulder's house, waiting to kill him too. Before Mulder can shoot the assassin, however, exacting the revenge he desires, Scully shoots Mulder in the shoulder, worried that his rogue status with the FBI will get him arrested for murder. Unable to either hear his father's story or avenge his death, Mulder is placed in a precarious position: the offspring of a member of the ultra-white male consortium he seeks to expose, Mulder must rectify his father's, and by implication his own, relationship to this bastion of oppressive whiteness.

When Mulder awakens after being shot, the elderly Navajo man Albert Hosteen is standing over him. Scully has driven them to New Mexico for aid in decoding their encrypted computer file. Scully explains to a confused Mulder that Hosteen was a Navajo code talker during the war, and that she learned of him from a friend in Washington, D.C. Although this friend got Scully in contact with Hosteen, however, Scully adds, "He claims he knew you were coming." "Last night we had an omen," explains Hosteen cryptically, beginning to incorporate Mulder into the Navajo narrative of the exegesis. Later, Hosteen takes Mulder to the reservation to show him what had been found there. As they drive up to Hosteen's house, Mulder inquires of Hosteen, "You said you knew I was coming." "In the desert, things find a way to survive," Hosteen answers, adding that, "Secrets are like this too. They push their way up through the sands of deception so men can know them." "Why me?" Mulder asks, to which Hosteen replies:

> You are prepared to accept the truth, aren't you, to sacrifice yourself to it?... There was a tribe of Indians who lived here more than 600 years ago. Their name was Anasazi. It means the ancient aliens. No evidence of their fate exits. Historians say they disappeared without a trace. They say that because they will not sacrifice themselves to the truth.

"You think they were abducted," Mulder quickly catches on. "By visitors who come here still," Hosteen adds. Framing Mulder as someone who is willing to sacrifice himself for the truth, Hosteen's narrative begins to separate Mulder from the white consortium in which his father participated, suggesting his existence outside this context of oppressive whiteness.

Unlike the writers of (white) history, who are unwilling to face the truth, Mulder is willing to sacrifice himself to know what is buried in the desert sands of the reservation. Following his conversation with Hosteen, he goes with the young Navajo man Eric to see the things he has uncovered in the desert. As Mulder nears their destination, his cellular phone rings, the anonymous Cigarette Smoking Man calling to urge Mulder not to continue. "Expose anything and you only expose your father," the Cigarette Smoking Man threatens. Willing to sacrifice himself and his father for the truth, however, Mulder continues, vehement with the Cigarette Smoking Man that he will uncover what has been hidden. When Mulder sees the object in the ground, he brushes the dirt away, exposing a plaque that reads "Sierra Pacific Railroad." The object is an old railroad boxcar, a refrigeration car that has been buried in the sand. Climbing inside, he sees deformed corpses stacked throughout the boxcar. From inside he calls Scully, telling her what he has found. But Scully has news for him as well. Having gone through the encrypted files with Hosteen, she explains:

In these files I found references to experiments that were conducted here in the U.S. by Axis power scientists who were given amnesty after the war. Some kind of tests on humans, what they refer to as merchandise.

While Mulder initially believes these deformed bodies to be alien corpses, when he looks closely he sees that they have smallpox vaccination scars—suggesting that the bodies are human. Apparently evidence of these human experiments performed during and after World War II, these corpses further "whiten" the consortium to which Mulder's father belonged, suggesting their similarity with and connection to the Nazis, the whitest group of white oppressors in memory. Mulder's search for the truth has exposed his father's oppressive whiteness still further, further problematizing Mulder's own identity.

As if in reply, the Cigarette Smoking Man and a troop of army soldiers show up in helicopters. Eric closes the boxcar with Mulder inside, hoping to protect him from being captured. "What's your name, boy?!" the Cigarette Smoking Man asks Eric, his emphatic "boy" emphasizing his racism and further highlighting his ultra-white whiteness. "Where's Mulder?" the Cigarette Smoking Man asks again, though Eric remains silent, unwilling to give Mulder up to these federal men. The soldier's climb inside the boxcar, but they cannot find Mulder. The Cigarette Smoking Man orders the soldiers to burn the boxcar of corpses, hoping that Mulder is inside. Another hardly subtle reference to Nazi era Germany, the burning boxcar illustrates the Cigarette Smoking Man's hyperwhiteness even more. Leaving Mulder for dead, the federal men take Eric and fly away in their helicopter (thus ending season two). Again, like the hyperwhite Borg, this consortium is cast as an unthinking mass of nameless bodies, willing to hurt anyone who might get in their path.

The next set of scenes, which narrate Mulder's survival, pass him from the world of whiteness to the world of the Navajo, relieving him of his connections to the oppressiveness of white history and his genealogical tie to his father's role within it. The scene opens with Albert Hosteen's voiceover, a narration that makes explicit this battle over history and Mulder's righteous place within it.

There is an ancient Indian saying that something only lives as long as the last person who remembers it. My people have come to trust memory over history. Memory like fire is radiant and immutable while history serves only those who seek to control it, those who would douse the flame of memory to

put out the dangerous fire of truth. Beware these men for they are dangerous themselves, and unwise. Their false history is written in the blood of
those who might remember and those who seek the truth.

The X-Files makes clear who is who in this battle between those who
seek the truth and those who write false history in their blood. As
Hosteen's narration finishes, a group of military men come to "question" him and his family, leaving them bruised and beaten. In contrast, Mulder is seen as a seeker of truth in commune with the spirits
of the Navajo. After finding Mulder, barely alive, burrowed in a hole
in the ground, Hosteen narrates again:

The desert does not forgive man his weakness. Weak or strong it takes no
mercy, and can kill a man in less than a day. To survive one must develop
skin like leather, know where to find water and when to take shelter. The
FBI man would have surely died had he not stayed underground, protected
like the jackrabbit or the fox. Even so, death was near.

Identifying Mulder with the jackrabbit and fox, and commenting on
his ability to survive the desert, Hosteen's narration frames Mulder as
a Native American, in contrast to the values of whiteness demonstrated by the Cigarette Smoking Man and the other men with whom
his father worked.

But Mulder's real passage beyond whiteness happens during the
"Blessing Way Chant," the ritual that Hosteen and his fellow Navajo
perform to help Mulder recover from his time in the desert. "This
healing ritual called the Blessing Way has been passed down by our
ancient Navajo ancestors," Hosteen explains in a voiceover as he prepares Mulder for the ritual, painting designs in the sand and covering
Mulder's body in evergreen branches. Calling the "Holy People" to
come and heal Mulder, the ritual is intended to bring him back from
the brink of death. Mulder's struggle, however, as Hosteen narrates it,
is not merely escaping death but breaking himself away from his ancestors as well.

My fear for the FBI man was that his spirit did not want to be healed. That it
wished to join the spirit of his own father who had died and did not want to
return to the world of the living. If the struggle to continue is too large or
the wish to join the ancestors too strong, the body will give up. But if the desire to resume life burns bright enough, the Holy People will be merciful.
The days and night now will be long and difficult for the FBI man as the
Holy People come and help him to choose.

While Hosteen's narrates, the other Navajo men beat drums and chant over Mulder's body, laid on a log bed and covered in branches. Floating before a star-filled sky, Mulder seems to leave Earth. Here, he is confronted with spirits from his past, including that of his father. "I stand here ashamed of the choices I made when you were just a boy," his father tells him, reiterating his own connection to the ultra-white consortium of government agents. "You are the memory, Fox. It lives in you." His father continues, "If you were to die now, the truth would die and only the lies would survive." As if these fatherly words were all he needed to "make his choice," Mulder leaves his white ancestors and returns to the living world of the Navajo reservation. Blessed by the Navajo Holy People, Mulder is newly remade, separated from this past of whiteness. As Hosteen explains of Mulder's transformation, "Like a rising sun, I sensed in him a re-birth."

Reborn via this Native American ritual, Mulder is free "to seek the truth." As the story continues, he and Scully continue their quest to expose the consortium's activities, exposing still further ties to the Nazis and a warehouse filled with the medical records of people the consortium's scientists have subjected to various sorts of genetic testing. And yet, for all the talk of truth seeking and memory, *The X-Files* fairly easily leaves this story behind, happy to push it into the past and continue their metanarrative. After all, "solving" this problem, finding "the truth" at the beginning of the third season, would effectively end the show's story line, solving the mystery that drives it. Instead, the mysterious stolen tape is returned to the consortium's members, and Albert Hosteen's memory of it, passed orally to twenty other Navajo men as well, is used to blackmail the bureau into reinstating Mulder and Scully and agreeing not to have them killed. The tapes and their contents will be forgotten unless there is a future need to remember them. The consortium is protected for the time being, but Mulder and Scully will presumably be back finding new ways to expose their practices. Crisis averted for the time being.

Mulder's "rebirth" and his "rejection of his ancestry" serve a like feat of selective memory. Like Captain Picard's struggle with the Borg, which works to unwhiten Picard and the Federation via the glare of the ultra-white Borg, so the glaring, neo-Nazi whiteness of the consortium effectively unwhitens Mulder, as does his ritual induction into Navajo culture. While embracing a conspiratorial perspective that frames this group of high-ranking state officials as evil oppressors may serve an important kind of challenge to authority, encouraging skepticism about federal agencies and their "good will," these concomitant visions of whiteness are also problematic in several regards. The caricature of hyperwhiteness associated with the consortium, for instance, may serve to wash away the perception of other, more subtle invocations of white hegemony, particularly the visions implicit within the principal characters of *The X-Files*, the characters with whom the American audience is supposed to identify. Like John Dunbar in *Dances with Wolves*, Fox Mulder is given as the better Native American, the white man the Navajo rescue to save their memory and to find their truth.

And yet, as *The X-Files* frames it, "the truth" is too messy and contingent to ever really be located anyway. The program's motto, "the truth is out there," restated at the beginning of every episode, is not merely an observation but a demand. As illustrated in these three episodes, the narrative of *The X-Files* demands that truth always be just out of reach, that it be muddied just enough to ensure another season of truth seeking. Likewise, the show's "truth" vis-à-vis whiteness is just as muddled, rescuing a mainstream American whiteness at the same time that a hyperbolic, neo-Nazi whiteness is condemned. Mulder's newly rescued whiteness, like that of John Dunbar's, depends on his capacity to speak for the Navajo people, like Dunbar for the Sioux. And still more so than Dunbar, Mulder's rescued whiteness depends upon his chaotic nothingness, the chaos of competing stories and stereotypes, to unidentify his character. Explaining Mulder's, recovery, Albert Hosteen draws upon a Native American myth:

> When the FBI man Mulder was cured by the Holy People, we were reminded of the story of the Gila Monster, who symbolizes the healing power of the medicine man. In this myth, the Gila Monster restores a man by taking all of his parts and putting them back together.

The X-Files effectively takes apart the history of whiteness—with Mulder's body—and rebuilds it, remaking it into its unblushing, nothing self.

Whiteness in Lethal Weapon IV

Discussing the late '80s police film *Lethal Weapon* (1987), the original film in the series starring the white Martin Riggs (Mel Gibson) and African American Roger Murtaugh (Danny Glover), film and media scholars have commented on the film's interesting and problematic narratives of interracial bonding. While the black and white buddy narrative has a long history in American fiction, stretching back at least as far as Huck Finn and fugitive slave Jim, critics have noted unique class dynamics within *Lethal Weapon*, a dynamic that seems to invert stereotypes of white and black "success." Contrary to common narratives of the poor black man and the malevolent white male who helps him (seen, for instance, in Atticus Finch's altruistic legal defense of African American Tom Robinson in *To Kill a Mockingbird*), Murtaugh is here the successful, established male who must lend a leg up to the troubled, white Riggs. While the recently widowed Riggs lives alone in a trailer, where he contemplates suicide and seems to flash back to his days in the Army's Special Forces division, Murtaugh lives a happy bourgeois life, complete with wife, family, and appropriately middle-class appointed home.

But while this could be read as a positive portrayal of a black male masculinity that escapes the traditional stereotypes of the past, both Robyn Wiegman (1995) and Fred Pfeil (1995) warn against such celebrations. As Wiegman suggests, *Lethal Weapon* still frames the white Riggs as the hero of the film, the protagonist who eventually rescues himself, Murtaugh, and Murtaugh's daughter from the hands of vicious terrorists. Unlike Atticus, whose superior bourgeois sensibilities work in rescuing Tom Robinson's character, Wiegman maintains that it is Riggs' violations of bourgeois sensibilities, his relentless insanity in battle and disregard for his own life, that allow for his heroism. In an odd reversal of Atticus and Tom, it is precisely Riggs' opposition to bourgeois civility that allows him to rescue Murtaugh's bourgeois life, returning him and his family to their comfortable, middle-class existence. Like the nonbourgeois Tom who needs Atticus to fight on his behalf, the bourgeois Murtaugh needs Riggs to do the same. Thus, while a bourgeois sensibility can be framed as heroic on the body of the white Atticus, it works differently when applied to Murtaugh. As Wiegman argues:

> The bourgeois black male, supporter and defender of the family, ensconced
> in heterosexuality, and in debt (psychically, if not financially) to the heroic
> white male offers a scenario of cultural acceptance that not only reinscribes
> the supremacy of father/family/capital/heterosexuality but also implicitly
> criminalizes the categories of opposition on which these images of cultural
> centrality and normativity dwell. In this regard, the gleaming positivity of
> black men in interracial male configurations serves as both the precondition
> for and symbolic achievement of integration's possibility, a possibility ar-
> ticulated within the normative categories instantiating the subject in U.S.
> culture: masculinity, heterosexuality, and Anglo-American ethnicity. (pp.
> 145–146)

Riggs and Murtaugh's "role reversal" is thus no even trade but a re-
distribution of bourgeois masculinity that allows for the rearticulation
of heroic white masculinity. In an '80s climate of hard-bodied,
Reaganomic masculinity, *Lethal Weapon* dispatches a bourgeois sensibility to integrate Murtaugh within the confines of whiteness, freeing
Riggs to exercise an unbridled, hard-bodied, heroic maleness, Riggs'
friendship with Murtaugh connecting him to bourgeois life without
forcing him to live it. In the process, "whiteness" is rescued both in
spirit, as Murtaugh's "success" demonstrates, as well in body, the
toughened white male body the heroic site of struggle and victory.

The late '90s film *Lethal Weapon IV* (1998), in contrast, offers a new
vision of whiteness and masculinity, one in conversation with both
the '90s crisis of masculinity and the growing discourses of multiculturalism and political correctness that have proliferated throughout
the '90s as well. Leaving behind the violent, hyperaggressive, borderline insane Riggs of the first *Lethal Weapon*, this later film depicts a less
chaotic, more lighthearted vision of masculinity in Riggs' character.
While the Riggs of *Lethal Weapon* was a loner, on the margins of the
police force, through several episodes of character development, he
has transformed into a hilarious policeman-comic who keeps his fellow officers in stitches with his zany antics and pranks. For instance,
Riggs regularly posts embarrassing newspaper articles about his
partner, Murtaugh, and then feigns disgust at his "fellow officer's
childishness" when they laugh, the other officers all the while knowing that Riggs himself is the trickster behind the prank. The gradual
inclusion of comic relief Joe Pesci and, for *Lethal Weapon IV*, comedian
Chris Rock, highlights this comic element still further. Likewise, involving Riggs in a monogamous relationship, these later *Lethal
Weapon* films temper his previously hyperviolent, psychotic characteristics with a narrative of romantic love and, eventually, family. Shift-

ing the *Lethal Weapon* narrative and Riggs' character in these ways, *Lethal Weapon IV* whitens its white hero for a new decade, wresting the comforts of middle-class existence back upon the body of white male heroics.

The opening sequences of *Lethal Weapon IV* demonstrate this change in Riggs' character clearly. Called to deal with a terrorist cloaked in armor and firing upon the streets with a blow torch and machine gun, Riggs and Murtaugh must find a way to defeat this bizarre criminal. "Don't turn around. Don't turn around. Don't turn around," Riggs chants as they drive their car towards the terrorist, comically "willing" him not to shoot them. Whereas the younger Riggs would presumably have driven straight into the melee, illustrating his insane heroics and disregard for his own life, this later Riggs resorts to superstitious chanting to protect himself and his partner. Riggs' final plan for defeating this villain is equally comical. "Here's what were gonna do," Riggs says to Murtaugh, "Take your clothes off." "What the hell for!?" Murtaugh replies, confused. "What for," Riggs thinks, before explaining, "Okay, you run, flame-o here turns, sees you in your undies. It distracts him, I know it would distract me, and that's when I shoot him." After Murtaugh takes off his clothes, Riggs adds, "Also, flap your arms and make noises like a bird…. It'll distract him more." Murtaugh does as Riggs suggests. Flapping his arms and squawking, Murtaugh runs in front of the terrorist, distracting him just long enough for Riggs to shoot him. "Hey Riggs, do you think that the bird thing helped?" Murtaugh later asks enthusiastically, proud of his role in defeating the terrorist. "No, no," answers Riggs, laughing, "I just wanted to see if you'd do it." "Cute shorts, Rog," Riggs adds as they walk away. A thorough trickster, Riggs now cares more about playing jokes than defeating criminals, a far cry from his hyperviolent, marginalized character in the first *Lethal Weapon*.

During the course of this comic mayhem, we learn that Lorna, Riggs' live-in love interest, is pregnant, and that Riggs will soon be a father. Also, we find out that Murtaugh's daughter Rianne is pregnant and will soon make Murtaugh a grandfather. These two narratives help ground the tale of family that runs throughout *Lethal Weapon IV*, creating the organizing sensibility around which Riggs' whiteness is remade. After the defeat of the terrorist in these opening scenes, the film begins again "almost nine months later" (as the subti-

tle explains), Riggs' impending fatherhood a more immediate reality
for his character. Sitting together on Murtaugh's boat, Murtaugh confronts Riggs. "You know, I've kept my nose out of it," Murtaugh begins. "You and Lorna, you're having a baby ... you live together ...
you're not married." "Does that violate your family values system or
something?" Riggs asks. To which Murtaugh responds, "I'm just trying to understand." In response, Riggs retells the story of his former
wife:

> You know, when I was married, you know, it was good. You put the ring
> on, you make the commitment, you do all of that. She's dead cause I'm a
> cop. She's dead cause I'm a cop, right. I mean, I don't wear the ring any
> more, but, you know, I look down, it's still there. I still feel it and, I don't
> know, it's like I'm not finished with that.

While the younger Riggs seemed marginal to the bourgeois family life
of Murtaugh (as Wiegman, above, suggests), this elder Riggs seems to
understand it all too well. Seemingly trying to deal with his memories
of his part marriage, Riggs seems to be working towards reconcilia-
tion with his bourgeois, married past.

The story of Murtaugh's impending grandfatherhood proves a
similarly important articulation of bourgeois family values. When
Murtaugh first learns that Rianne is pregnant, he replies, "But Rianne
can't be pregnant, she's not married." His bourgeois sense of family
values unable to process Rianne's pregnancy, Murtaugh reacts in confusion and frustration. Murtaugh's lecture to Riggs about having a
baby but not getting married is hence as much for his own daughter
as for Riggs. Here, regarding Rianne's pregnancy, and thus Murtaugh's bourgeois family, the film traverses a strange slope of racial
stereotypes. Rianne's pregnancy seems an invocation of a traditional
stereotype of black female reproduction—the hyper-reproductive,
out-of-control welfare mother an all too common trope in American
discourses of race and reproduction. And yet, the film works to push
this stereotype away as well. We soon learn that, unbeknownst to
Murtaugh, Rianne has secretly married the soon-to-be father of her
child, African American police officer, Lee Butter (Chris Rock). Knowing that Murtaugh would be angry if he knew his daughter had married a cop, Rianne, her family, and Riggs all keep this marriage a
secret from him. As a result of this secrecy, for Murtaugh, Officer Butter is simply the weird new young officer on the force, the anonymous cop whose name he and Riggs constantly misspeak. In fact,

Murtaugh regularly confuses Butter's attempts at friendship for homoerotic desire, an ironic misrecognition, the film implies, given Butter's relationship with Murtaugh's daughter. This "misrecognition"—of Rianne as unmarried, of Butter as homosexual—ultimately works to reinforce a sense of bourgeois normativity, the secret bourgeois family values underwriting Rianne and Butter's relationship.

In this way, Rianne's pregnancy and relationship with Butter both invokes and pushes away a series of stereotypes of black sexuality and reproduction. In that Rianne's sexual partner is unknown to her father, despite the fact that they live in the same town and work on the same police force, Rianne and Butter's relationship is the stereotype of black hyper-reproduction—nearly anonymous partners who reproduce without any thought of the consequences (for instance, the alienation of Rianne's father, Murtaugh). Likewise, Murtaugh's misrecognition of Butter's nervous friendship as homosexual desire reiterates the hypersexuality of the black male still further. Fully sexualized, what could Butter's smile be, but a smile of sexual longing? At the same time, by evoking the trope of bourgeois marriage, *Lethal Weapon IV* ultimately seeks to salve this hyper-reproduction and hypersexuality. "Knowing" that Rianne and Butter are "really married," the viewers, like Riggs and the others of the film, share the inside joke that makes Murtaugh's frustrations and confusions the comical moments that they are. With a happy, bourgeois marriage posited as the punch line to these comedic moments, these bourgeois family values are rescued once again within the film's exegesis. And yet, by doing so only through the evocation of these tropes of black hyper-reproduction, the film likewise demonstrates the racial costs of its celebration of bourgeois normalcy. Only by evoking a stereotypical blackness, these incidents suggest, can *Lethal Weapon IV* hold up these (white) bourgeois family values to celebration.

Still more clearly, the sense of family demonstrated within the villains of the film further evinces this manner of celebrating bourgeois normalcy at the expense of racial otherness. Its plot centered on an Asian criminal's attempt to free his brother from prison, and thus to unite their crime family, the film offers up a bizarre picture of Asian familial relationship. Whereas the good, whitened Americans of the film evidence a healthy, bourgeois familial devotion, the film's Asian villains evoke a dedicated sense of family that is depicted as evil nonetheless—the mirror against which bourgeois family values are reflected and magnified. In his attempt to free his brother and his brother's partners from prison, the villainous Wah Sing Wu (Jet Li), has "commissioned" a Chinese artist to counterfeit money with which to bribe the Chinese general holding these men prisoner. In "payment" for the artist's service, Wu has made arrangements to smuggle the artist's family in from China and to provide them papers so they may live in America. When the ship and its contents are hijacked by Riggs and Murtaugh, Wu begins a search for the artist's family, the Hongs, hoping to find them in time to motivate the artist to finish his work and to buy back his brother with his newly counterfeited money.

Wu's relentless attempts to unite with his brother are cast not as genuine attempts at family togetherness but rather as selfish, misguided, and violent disregards for family more generally. When Wu learns that Sergeant Murtaugh has hidden the Hongs at his house (further evidence of Murtaugh's own relentless dedication to bourgeois family values), he takes a group of men to retrieve them. Arriving while Murtaugh is gone, Wu and his henchmen capture the Hong family, Murtaugh's wife and daughter, and Riggs' partner Lorna, who is visiting. The way in which Wu et al. treat these families positions these Asian villains as the antithesis of bourgeois family values, as the opposites of the heroic family values of the film. Riggs and Murtaugh show up while these villains are there, pull their guns on Wu and his men, and free their families. Demonstrating their antipathy towards American, bourgeois family values, however, one of Wu's men pulls a knife and holds it to Lorna's pregnant belly, threatening to kill Riggs' progeny if Riggs and Murtaugh don't drop their guns. Of course, Riggs and Murtaugh comply, their own dedication to bourgeois family showing once again. In the end, Wu and these other villains tie up Riggs, Murtaugh, and their families and light their house on fire, leaving them to burn.

In this rejection of the bourgeois family, Wu and the other villains serve to heighten the validity that *Lethal Weapon IV* accords to this middle-class American family unit. Working as a foil to middle-class family values, the values of family attributed to Wu are presented as dangerous and wrong—a parody of real (white, American, middle-class) family love. A misunderstanding between Riggs and Murtaugh and these Asian mobsters demonstrates this clearly. Riggs and Murtaugh originally hear these villains' planning to unite with their "forefathers," believing that they have some goal of joining their ancestors in some newly established familial bond. However, they later learn correctly that the plan is to unite with the "Four Fathers," the group of four crime bosses that includes Wu's brother. Here, a seemingly genuine familial bond is turned on its head, exposed as a mere selfish ploy to create a powerful crime syndicate. In the end, Riggs and Murtaugh are able to stop Wu from buying back these Four Fathers, exposing their counterfeiting scheme to the general they attempt to bribe. In the ensuing fight, Murtaugh inadvertently shoots Wu's brother, fatally wounding him in the chest. Even with his own brother dying in his arms, however, Wu's grief is still presented as a parody of real (white, American, bourgeois) family love. Mocked by the comic jabs of Riggs and Murtaugh, Wu's brotherly love is given as an extension of his hyperaggressive anger and villainy: "Now you've done it," Riggs says comically. "Now he's really pissed." In a vicious fight that leaves both Riggs and Murtaugh wounded, they are somehow able to kill Wu, despite his deadly martial arts expertise. Cast as the heroes of this battle, Riggs and Murtaugh, who pierce Wu through with a sharpened metal rod, seem the heroic champions of real family bonds, destroying Wu so as to avenge the wrongs done to their own family units.

This defeat of Wu, and thus the anti-American family values the film associates with him, is important to Murtaugh's character in another regard as well. Cast as an aging, borderline over-the-hill cop experiencing difficulty with his job, Murtaugh needs this final battle to prove that his masculinity is still intact, that despite his age and newfound gentility, he can still get the job done. An earlier scene has Riggs sparring with a younger officer in a boxing ring. Wearing head gear and boxing gloves, the movie seems to have domesticated the aggressive physicality associated with Riggs' character—this ex-Special Forces soldier who, at the end of the first *Lethal Weapon*, fights a different villain hand-to-hand in the rain ("Would you like a shot at the title?" this younger Riggs' says to the villain, egging him into this confrontation). This older Riggs, however, fakes a shoulder injury in order to get out of this boxing match, distraught over his aging body and his inability to exercise his youthful skills in the martial arts. "I hate to say this," Riggs tells Murtaugh, resting in the locker room. "You're getting too old for this shit," Murtaugh replies, finishing Riggs' sentence. "What am I gonna do?" Riggs asks Murtaugh. Answering him, Murtaugh responds in a matter-of-fact voice, "Accept it. Like I do."

Riggs' aging, changing body works analogously to his changing character more generally. Unable to sustain the image of the '80s hard-body cop that launched the *Lethal Weapon* series, Riggs' character must accommodate a new decade of masculinity. But whereas *The X-Files* accommodates Agent Mulder to this new era of masculinity through an "unwhitening" of its main character that dissolves his identity in various ways, *Lethal Weapon IV* rescues Riggs by taking shelter in the visions of bourgeois family values presented throughout the film's history, though previously associated with Murtaugh rather than Riggs. And yet, maintaining the new, comic, trickster elements of Riggs' character, even this is an uneasy fit—a joke slid into haphazardly. Finally deciding he is ready to marry, Riggs proposes to Lorna as she is wheeled into the delivery room on her gurney. "Married" by a rabbi outside the delivery room, the two enter into the bonds of bourgeois marriage before their child is born, assuring the film's middle-class family values will remain intact. With this new, happy, family unit created, and sprinkled with the film's developing sense of humor, the ending of *Lethal Weapon IV* seems more situation comedy than cop thriller. At the hospital, with Lorna, Riggs, and their baby, and Murtaugh, Rianne, and her baby, as well as the rest of the Murtaugh family, all posing together for a picture, the film encapsulates photographically the bourgeois unit it has celebrated with the film. "Are you all friends?" the doctor who takes the picture asks before snapping the camera. "No, we're family!" they all answer in unison.

This model of bourgeois family created, Riggs' masculinity is rescued and remade for the later '90s. Celebrating his newfound bourgeois gentility, though with a comic edge, the film works to break Riggs still further from his hard-body past, casting him as a new sitcom dad, all happiness and smiles and touched by bourgeois family values. In the process, the film offers a newly whitened vision of the bourgeois family. Playing off of the narrative of Rianne's pregnancy—a story of black hyper-reproduction whose release is bourgeois normalcy—as well as the story of Wu's selfish, violent bonds with his brother, the film creates a safe space of white bourgeois family values in which this elder Riggs may dwell. One of the film's promotional pictures makes this relationship particularly clear. In it, Mel Gibson (Riggs) faces the camera, taking up the majority of the left side of the frame. Behind and to the right is Danny Glover (Murtaugh), slightly in profile and smaller than Gibson. In a line from behind Glover, each slightly smaller than the one before, are Joe Pesci (Leo Getz), Chris Rock (Lee Butter), and Jet Li (Wa Sing Wu), with Rene Russo (Lorna Cole) slightly behind Gibson. Given as Mel Gibson's supporting cast—the slightly smaller faces who stand behind the great white hero—these multicultural cast members are the others through which Riggs' heroic masculinity is won, the others to this lethal weapon of a newly unblushing, white masculinity in America.

Presidential Whiteness and the Performance of Identity

Together, these '90s heroes rearticulate a sense of the no-thing-ness of white. While the sort of fluidity of identity associated with Bill Clinton's ability to be read as the first black president might suggest the possibility for highlighting the fictiveness of whiteness, and thus dislocating it, in the above examples this fluidity works to reinstate an unblushing American male, recreating the abstract, unblushing citizen associated with dominant models of masculinity. John Sloop (2000) recently responded to these notions of the "performativity" of gender. Discussing Brandon Teena, the transgendered man depicted in *Boys Don't Cry*, Sloop argues that notions of performance are in many ways ill equipped to deal with the specifics of Brandon Teena's life and death. As he explains:

> As has become commonplace in many contemporary discussions of gender and sexuality, at least since the publication of Judith Butler's *Gender Trouble*, gender and sexuality are assumed in this essay to be potentially fluid, held in check by each individual's interpellation into a cultural ideology that maintains male-female differences.... [And yet,] while this move to celebrate or highlight potential disruptions of the gender binary system is indeed a vital project, it can come at the cost of focusing on ways that the dominant rhetoric/discourse of gender continues to ideologically constrain. (pp. 167–168)

Continuing, Sloop argues that: "while cases of gender ambiguity obviously have the potential to cause 'gender trouble' and disrupt bigender normativity, in terms of the 'dominant' discussions that surround such cases ... the 'loosening of gender binarisms' is a potential that often goes unrealized for many audiences" (p. 168). The case of Brandon Teena's death in particular, Sloop argues, illustrates that "many institutions and individuals work to stabilize sex, to reiterate sexual norms, rather than to encourage/explore gender fluidity" (p. 168).

While this is an interesting and important caveat to celebrants of gender fluidity, it overlooks a second caveat, the next step of analysis demonstrated within this chapter (and, indeed, throughout the preceding chapters as well). As illustrated in the case studies above, the encouragement of a kind of "gender fluidity" may itself work to reestablish dominant figures of identity, though often in different guises and with different ways of operating. The character of Andy Kaufman in *Man on the Moon* is whitened precisely through the performativity of his identity, that is, precisely through his ability to be "fluid." It is through his multiple performances, his multiple and competing identities, that Kaufman's identity is washed away, leaving the film with the impression that "there is no real Andy Kaufman." But Kaufman's performativity does not so much liberate us into a realm of utopian fluidity as offer up a no-thing-ness that whitens Kaufman to the point of forgetting his Jewish identity completely. With such examples in mind, we should be as cautious about "potentials" of gender fluidity "that go realized," as with those that don't, always keeping in mind the context and conditions through which gender and identity more generally are made fluid.

In terms of Bill Clinton's "fluid" identity, these dilemmas are equally evident. While the fluidity of Clinton's racial markings may signal a kind of "possibility" or "potential" for upsetting racial standards and normativity, it opens up other sorts of possibilities as well, including that possibility for reinscribing a sort of whiteness against the stereotyped identification of some other. In Al Gore's and Bill Bradley's attempts to cast themselves as "the second black President" in hopes of securing a Democratic nomination (as evidenced in their February 2000 debate at the Apollo theater), they seemingly sought to stabilize the presumably floating blackness of which Clinton has partaken. Each attempting to show that he was blacker than the other, they worked to quantify their own racial identifications relative to the other's. Here, the "possibility" of Clinton's racial mobility is made a marketing strategy, a strategy that seeks not to disrupt, but to fix, hoping to call an audience of black voters into existence and thus to win their support. Not that Clinton's own racial mobility did not partake of the same marketing strategies. But the "possibilities" that Clinton may have opened up in the process seem to have potentially stabilized this strategy still further.

In the conclusion of her discussion of *Lethal Weapon*, Robyn Wiegman offers the following interpretation of the film's concomitant vision of white masculinity:

> In our seduction into the visual realm of culture, in our desire to find the visible liberating, in our subjective need to reclaim bodies from their abjection and recuperation, we encounter both the threat and the utopic possibility of contemporary social critique. To forge this critique as part of undoing the illusion of "man," of refusing the myth of masculinity's completion in an interracial domain devoid of women, is to expose the continuing rift within whiteness, that materially violent abstraction that perversely gathers strength by offering itself now as a struggling, innocent, and singular voice. (1995, p. 146)

To this we might add a whiteness empowered through multiple voices, a new sort of abstraction attained through fluidity and a seemingly schizophrenic putting on of identities. Negotiated, refracted, and rearticulated, these changing identity types are productive moments of possibilities; which is to say, they make possible the dissolution of dominant identities as well as their rearticulation via new and changing forms.

Notes

1. For further discussion of Clinton's potential blackness, see Deem (1999, 2001).

2. For further discussions of "whiteness," see Cuomo and Hall (1999), Nakayama and Martin (1998), Hill (1997), Delgado and Stefancic (1997), and Frankenberg (1997). Also see Stecopoulos and Uebel (1997) on the topic of race and masculinity. Arguing for an attention to whiteness, Uebel (1997) argues that an "attention to the specific historicity and textuality of privileged, often ideologically invisible, categories such as whiteness prevents the acceptance of their uniformity and autonomy" (p. 2).

3. Maintaining his arguments about whiteness, Dyer suggests that *Dances with Wolves* portrays the Sioux as "whiter" than the Pawnee, an indication of the ways

in which the goodness of whiteness still manifests itself in this presumable critique of whiteness and its history (1997, p. 60).

4. For a reaction to Goldman's conception of Elvis, see Marcus (1991) (especially pp. 47–59). Likewise, for a discussion of Elvis' contemporary manifestations and problematics, see Rodman (1996).

5. Indeed, although Kaufman's Jewish identity is all but expunged from *Man on the Moon*, references to his potential Christianity are relatively abundant, including a scene from Kaufman's performance at Carnegie Hall in which Santa Claus comes to spread happiness to the audience, and a mock Christian wedding that was televised on the program *Fridays* (though this wedding scene was included in the final script but removed in editing the final version of the film).

6. Joseph Stiglitz (April 17 and 24, 2000), former chief economist and vice president of the World Bank tells a frightening story of the IMF that sounds remarkably similar to the Borg of *First Contact*. Prior to the year 2000 meeting of IMF in Washington, D.C., Stiglitz wrote: "Next week's meeting of the International Monetary Fund will bring to Washington, D.C., many of the same demonstrators who trashed the World Trade Organization in Seattle last fall. They'll say the IMF is arrogant. They'll say the IMF doesn't really listen to the developing countries it is supposed to help. They'll say the IMF is secretive and insulated in democratic accountability.... And they'll have a point" (p. 56). Stiglitz likewise asserts that the "older men who staff the fund—and they are overwhelmingly older men—act as if they are shouldering Rudyard Kipling's white man's burden. IMF experts believe they are brighter, more educated, and less politically motivated than the economists in the countries they visit" (p. 57).

7. For a further discussion of the format and development of *Hill Street Blues*, see especially Gitlin (1983).

8. Discussing this division, and its concomitant bias against emotion, Williams and Bendelow (1998a), argue, "The roots of this neglect lie deeply buried in Western thought—a tradition that ever since Plato has sought to divorce body from mind, nature from culture, reason from emotion, and public from private. As a consequence, emotions have tended to be dismissed as private, 'irrational', inner sensations that have been tied, historically, to women's 'dangerous desires' and 'hysterical bodies.' Here, the dominant view seems to have been that emotions need to be 'tamed', 'harnessed' or 'driven out' by the steady hand of (male) reason" (p. 131).

Chapter 5

9/11 and After:

Masculinity, Citizenship, and National Crisis

In the November 19, 2001, issue of *Time* magazine, writer Lance Morrow attempted to describe what he saw as a new, post–9/11 America. In Morrow's estimation, the attacks of September 11 had brought the country into a new era, replacing the "Old Paradigm" of the pro-sensitive Clinton-era with the "New Paradigm" of a nationalistic, hypermasculine Bush age. According to Morrow:

> In the Old Paradigm, police were marginal blue blurs from the outer boroughs, and fire fighters merely the hired help. In the New Paradigm, they are the Heroes Who Rushed into the Burning Buildings When Everyone Else Was Running Out. (p. 152)

Tying this more explicitly to questions of masculinity, Morrow continues:

> In the O.P., machismo was a fault and the military an archaic and expensive nuisance. The N.P. admires strong men and manly virtues—courage and self-sacrifice. (p. 152)

In the era of the New Paradigm, Morrow asserts, "Bill Clinton is a relic of another time, like the 1920s party boy F. Scott Fitzgerald stranded in the landscape of the Great Depression."

For all of Morrow's talk of leaving the past behind, however, the images of September 11 depict a profound sense of masculine vulnerability that echoes and even amplifies the sense of masculine crisis at work in the '90s. From pictures of ashen-covered firefighters featured in U.S. magazines and newscasts following the attacks to descriptions of burned human bodies featured in the *New York Times* and elsewhere, the tragic aftermath of September 11 has been dramatized in the bodies of American citizens. The photographs in *Time* magazine's special September 11 issue clearly illustrate this profound sense of vulnerability and crisis, particularly in terms of the magazine's depiction of the September 11 male heroes. While the issue's pictures show the heroic firefighters and police officers of which Morrow speaks, hard at work to save lives, they also capture the tragic toll the events take upon these heroes. One photograph, taken by *Time* photographer James Nachtwey, shows four firemen, all wearing paper facemasks, standing or crouching on piles of building debris and discarded papers as they assemble a fire hose. The photograph's caption reiterates the conflicted sense of heroic action and tragic vulnerability captured within the photo: "New York's Bravest: Firefighters were still going in when buildings collapsed. One ducked under his truck and emerged to find everyone else in his squad dead."

Time's coverage here, like that of so many other news sources immediately after September 11, depicts a tragically conflicted masculinity, at once the embodiment of traditional masculine notions of strength and valor as well as the picture of pain and suffering. Another photograph in *Time*'s special September 11 issue illustrates this still more clearly. Taken by Ruth Fremson, the photograph shows a police officer resting in a deli near the World Trade Center. Covered in ashes and leaning with his back against the deserted deli counter, his head lowered, and his hands on his knees, this officer seems the perfect representation of the conflicted masculine heroics depicted in the immediate aftermath of the WTC attacks. The picture's caption reiterates this quite well: "While one cop caught his breath in a deli near the World Trade Center, another reported, 'I saw body parts all over the place, and I had to pull the car over to take a minute and cry.'" If the masculine hero of the '90s offered a conflicted blend of hypermasculine toughness and new age sensitivity, the September 11 hero is still more profoundly conflicted, eminently heroic and eminently vulnerable.

The conflicted notions of masculinity that inform these post–September 11 discussions illustrate the same conflicted possibilities that dominate the images of President Clinton and the host of '90s men explored in the previous chapters. President Clinton offered the '90s a conflicted masculinity characterized by ambiguous, mobile, and conflicted notions of sensitivity, sexuality, class, and race. Sensitive and tough, impotent and highly sexualized, Oxford educated and dirt poor, white trash and the first black president, President Clinton was a bundle of contradictions and conflicts. These conflicts allowed him to rewrite his character from moment to moment, helping him to maintain high popularity ratings despite accusations of waffling and repeated attacks against his character. At the same time, however, this same bundle of contradictions provided opportunities for Kenneth Starr's attack on Clinton, allowing Starr to frame Clinton simultaneously as too sensitive and not sensitive enough, as highly sexual and impotent. Here, the "possibilities" of Clinton's mobile identity were multifaceted, allowing for his heroism as well as his villainy. The conflicted images of firefighters and other September 11 figures alternatively seen acting heroically and breaking down in tears illustrates these same conflicted masculine possibilities, demonstrating the same profoundly troubled masculinity that characterized the '90s.

This notion of the multiple possibilities brought about by gender disruptions is the central problem informing this and the previous chapters. The "crisis of masculinity," in particular, can be seen to be productive in multiple ways. On the one hand, crisis of masculinity discourses have clearly offered up a series of challenges for American masculinity at the turn of the century. Working to mark the unmarked, abstract citizenship associated with Goffman's unblushing male in America, this proliferating discourse of crisis has worked to dislocate and decenter traditional notions of manhood. On the other hand, as I have attempted to show throughout, these discourses of crisis have also created the possibility for new and different formulations of this universal, abstract citizenry. Whether through a conflicted sensitivity, a co-optation of working-class cultural capital, or through a set of fluid racial markings that dissolve into whiteness, these dominant notions of American maleness find in this crisis talk escape routes through which to salvage the unblushing male.

The post–September 11 rhetoric deployed by popular magazines such as *Time* and *Newsweek*, in popular television coverage such as CBS's dramatic documentary *9/11*, as well as in repeated speeches and statements from the Bush administration, consistently draws upon and continues the same notions of masculinity that characterize the mid to late '90s. Indeed, a series of conflicted notions of masculinity characterize much of the discourses of healing, remembrance, and vindication that respond to the events of September 11. Here, masculinity becomes a key trope around which a variety of conceptions of the nation are built. For instance, the broken, defeated bodies of firemen and police officers stand as proof of our national vulnerability, fallen icons that demonstrate the profound tragedy that has befallen us. At the same time, the repeated discussions of these same firemen's and police officers' constant vigilance in rescuing people from the World Trade Center highlights a national masculinity that is still strong and intact. Similarly, repeated images of broken heroes suggest the disintegration of the American spirit, even as images and discussions of the American recovery effort seem to point towards a new, ultimate national unity.

This chapter considers a variety of post–9/11 messages, exploring some of the ways in which masculinity is performed and negotiated following the events of September 11. These various depictions of American manhood highlight the dramatic and often troubling linkages between masculinity and citizenship, making still more profound many of the central problems discussed in the previous chapters. Here, conflicted conceptions of masculinity combine with heroic celebrations of "manly virtues" (of the sort Lance Morrow speaks above), all alongside a celebration of new American unity that denies differences of race, class, and sexuality, much as white, abstract, masculine citizenship has traditionally done. In the midst of these conflicted negotiations of manhood, Goffman's unblushing, universal maleness returns with hyperbolic force, framing masculinity as the ultimate measure of nationhood, citizenship, and ethics more generally.

Homosocial Ritual and Masculine Crisis: CBS's *9/11*

Lance Morrow's desire to celebrate masculine heroics in the midst of a national crisis reflects an important trend within U.S. history. According to Michael Kimmel (1996), one of the country's earliest male heroes, the "self-made man," developed in the 18th century as the nation struggled to separate itself from the English homeland. Even after the revolution had been won, Kimmel suggests, American men found themselves struggling for identity, attempting to define themselves in opposition to the aristocratic notions of masculinity that dominated in England. The notion of the self-made man provided just such a masculine identity, allowing American men to define themselves through their hard work and ingenuity rather than through some aristocratic notion of birthright. Self-made masculinity was not tied to one's lineage (which, for most American men of the time, included ties to England) but rather to one's abilities to perform and make do in the new American public. As Kimmel explains:

> The central characteristic of being self-made was that the proving ground was the public sphere, especially the workplace. And the workplace was a man's world (and a native-born white man's world at that). If manhood could be proved, it had to be proved in the eyes of other men. From the early nineteenth century until the present day, most of men's relentless efforts to prove their manhood contain this core element of homosociality. From fathers and boyhood friends to our teachers, coworkers and bosses, it is the evaluative eyes of other men that are always upon us, watching, judging. (p. 26)

To be a self-made man was, and remains, about proving oneself alongside and against other men. These "homosocial relationships"—this sense of brotherly contact, cooperation, and competition—was important to early, mainstream conceptions of masculinity and remains so today, particularly during moments of crisis. From the Boy Scouts to bodybuilding, a variety of so-called masculine rituals have developed when the nation's manhood seemed threatened.[1]

The homosocial rituals of American men, like the masculine crises to which they often respond, lay bare a variety of contradictions underlying mainstream conceptions of American masculinity. On the one hand, as Kimmel suggests, homosocial behavior serves as *the arena* for demonstrating one's masculinity. From football to hunting and from drag racing to locker-room talk, American men are to prove their masculinity by both relating to and competing with other men in a variety of activities. The traditional bachelor party offers a strong example of this conception of homosociality. Not only are the activities of a bachelor party homosocial in that they typically involve men together doing stereotypically manly things but the entire event is meant to celebrate homosociality as a concept: the party presumably a lament for a brother who is passing into the feminine world of marriage and domesticity. At the same time, however, by placing men together in intimate contact, homosocial behavior challenges conceptions of mainstream masculinity even as it supports others. As an arena where men might be encouraged to share their deeper feelings, or come into intimate physical contact, homosociality plays against a variety of stereotypically masculine traits. For instance, some researchers have suggested that hypermasculine locker-room talk of one's sexual exploits may work to counter the potential nervousness men may feel from being in a room with other naked men (Curry, 1991). In this case, the homosocial space of the locker room seems to threaten the masculinity it seems intended to strengthen.

This conflicted sense of homosociality is clearly reflected in Robert Bly's work *Iron John* (1990), which I explored briefly in the first chapter. Bly's book took homosociality seriously, prompting a variety of new age men in search of their manhood to take to the woods in spiritual consciousness-raising sessions. Here, the book served as a celebration of male intimacy, encouraging men to both get in touch with their own feelings and to share their feelings with other men. In this sense, Bly's work stood as a challenge to conventional notions of American masculinity that have stereotypically encouraged men to avoid intimacy and to avoid expressing their feelings, at least in so blatant a manner. At the same time, however, Bly's book also reiterates a variety of traditional, hegemonic notions of manhood as well. In particular, Bly sees a particular "warrior masculinity" as an *essential feature* of the universe (not as a social or historical construction) reiterating the essentialist, universally generalizable characteristics that traditionally have given masculinity its hegemonic power.

> What I'm suggesting, then, is that every modern male has, lying at the bottom of his psyche, a large, primitive being covered with hair down to his feet. Making contact with this Wild Man is the step the Eighties male or the Nineties male has yet to take. (p. 6)

By focusing on this natural warrior masculinity, Bly seeks to rescue his homosociality from its potential contradictions, making clear that his is a very (traditionally) masculine homosocial bond. Similarly, the Promise Keepers have offered a call for more male intimacy, inviting men to join together to cry, hug, and to share their stories of masculinity, even as they reiterate very traditional ideas about the roles of women, sexuality, and religion. In both of these cases, a fairly radical call for homosocial ritual and intimacy seems accompanied by an equally extreme call for traditional notions of masculine authority and power.

This section explores homosociality as an element of the aftermath of September 11. Looking at the CBS documentary *9/11*, I investigate one way in which the popular media attempted to make sense of the horrific tragedy of September 11. This movie began as an attempt by two French filmmakers, brothers Gedeon and Jules Naudet, to tell the story of a rookie firefighter as he struggled through his first year of work with a New York firehouse—Engine 7, Ladder 1, the firehouse responsible for tending to the World Trade Center buildings. Following these firefighters into the WTC on September 11 the movie became a chronicle of one firehouse's experience on that tragic day, documenting both the profound fear and suffering of this group of men as well as their miraculous, heroic survival. Narrating this simultaneously as a masculine coming-of-age story for the rookie firefighter and as a story of the brotherhood of the firefighters as a whole, *9/11* captures and celebrates a complicated sense of homosociality, demonstrating some of the conflicted ways in which masculinity and masculine ritual serve as a particular kind of resource during moments of tragedy. This film caught the attention of the American public not only because it depicted the profound trauma of September 11 but also because it performed a sort of masculine impotence and survival that became part of the mythos of that tragedy more generally.

Bustin' Chops: Male Bonding and Masculine Initiation

Towards the opening of *9/11*, Gedeon Naudet, one of the two brothers who shot the film's original footage, discusses the original intent behind their documentary. "We wanted to show how a young kid becomes a man in nine months," he explains, referring to the initial probationary period that every new firefighter (or "probie") experiences. The next sequence shows a series of shaven-headed probies at work at the fire academy before the camera settles on Antonio (Tony) Benatatos. "It sounds kinda cheesy," Tony begins, facing the camera, "[but] I always kinda wanted to be a hero." Selecting Tony as the probie to follow, the film begins its masculine coming-of-age story, working to show how the other firemen who will become his peers gradually accept him into their brotherhood. In the process, the film highlights a series of challenges central to American masculinity in general.

At the beginning of the film, Tony's nerves serve as evidence of his green, probie status, illustrating the extent to which he has not yet been tested as a firefighter. "I'm terrified. This is what I want to do, but it's scary," Tony explains to the camera as he drives to his first day of work at the station. "I just hope I can do everything I'm sup-posed to do. I just don't know how I'm going to react when there's fire flying over my head." Arriving in the office of his new boss, Tony's nerves seem to get the best of him again, as he stumbles around in search of a chair. "Do you need to sit down?" one fire-fighter asks him somewhat jokingly. "Do you want to stop calling me sir?" he continues, teasing Tony for his overly formal attitude. "I'm a little bit nervous," Tony tells the room full of firefighters, as if that weren't already obvious to everyone watching. Tony's nervousness and the teasing by his fellow firefighters frame this opening series of images, setting the mood for Tony's initiation into firefighting and, the film suggests, manhood.

Further illustrating Tony's status as the low man on the totem pole, the next scenes depict Tony doing a variety of chores around the firehouse. "When you're a probie," James Hanlon, a senior firefighter who narrates much of the film explains, "what you're supposed to do is pretty much everything." Next the film cuts to a montage of Tony's various tasks: scrubbing pots in the sink, mopping the floor of the ga-rage, making beds, polishing the fire pole, getting instructions on the proper way to clean a fire truck. Tony's own voiceover during this sequence captures the monotony of his new rookie position clearly. "Probie rules," Tony begins, "Probie always gets in the sink [to do the dishers], Probie does not go in the TV room, ever, Probie makes sure there is always hot coffee in the morning." As a probationary fire-fighter, not yet initiated into the brotherhood and not yet having ex-perienced a real fire, Tony remains at the bottom of the pile. "I think I'm doing decent," Tony explains, "[but] I'm still waiting for a fire."

In addition to Tony's various chores, the film also captures the teasing that Tony undergoes as part of his initiation into the fire-house. "Tony wanted to prove to the other guys, and to himself, that he was going to be a great fireman," Gedeon Naudet narrates at one point early in the film. "The guys were not going to make it easy on him." In one scene, a senior firefighter confronts Tony about the con-dition of his shirt ("Do you have an iron at home?) as another teases him in front of the camera ("Probie in a lot of trouble, take 2"). An-other scene has a group of fellow firefighters throwing water on Tony from a second-story window. After Tony's first paycheck, James Han-lon comments: "You gotta be kidding me. You couldn't even buy a six-pack with that. Holy crap!" Summing up the goal of all of this teasing, a fellow firefighter explains to Tony: "We're gonna break your chops until you laugh about it, because that's how we do it. We'll tease you to death until you start laughing." Here, the film suggests along with these comments, we see Tony gradually being broken into the fellowship of the firehouse.

In a similar manner, the film illustrates the ways in which the filmmakers Jules and Gedeon are themselves made fun of and, the film suggests, accepted by the firefighters of the house. In one scene, the night of September 10th, Jules apparently decides to make a French dinner of leg of lamb for the firefighters, an attempt that doesn't go particularly well, as his brother Gedeon explains. "He cooked one and we really needed at least five," Gedeon laughs. The next scenes show the other firefighters eating the meal and poking fun at Jules' attempt. "Where's Frenchie?" one firefighter asks. "A couple more meals like that, we'll be able to share shirts." "We stayed up late, just telling stories and bustin' chops," Hanlon remembers of the evening. "Even though the guys were making fun of us because we didn't cook enough," Jules explains, addressing the camera directly, "we were having a great time. We were getting accepted." Like the probie Tony, the film suggests, so were Jules and Gedeon being initiated into the fellowship of the house.

Reiterating that this sense of fellowship is, indeed, central to the life of a firefighter, the film captures a host of bonding activities among the men of the firehouse. "You do your job, you risk your life to help people," one firefighter explains, "and with time you become part of the unique, extended family of the firehouse." As this firefighter narrates, the film offers a variety of scenes depicting this extended family at play, from a picnic with their families to joking around in the kitchen as they prepare dinner. "Tell me one other job where everyone sits down to dinner together every night," James Hanlon insists, reiterating the sort of homosociality that is central to the firehouse. Once you are allowed within this family, as Tony presumably will be, the film suggests, you are a part of a lifetime fellowship with an intense, brotherly bond.

In addition to putting up with teasing and performing a variety of chores, Tony must, of course, fight his first fire to begin to prove his ability as a firefighter. Early in the film, James Hanlon explains that there are "two kinds of probies:" "Black Clouds," who bring all of the fires of the city with them and "White Clouds," who don't bring any. "This kid was one very white cloud," Hanlon says of Tony. "It's been four weeks, five weeks, I think, something like that, and still no fire," Tony laments a bit later. "But it'll come. Probably when I'm asleep and not ready for it, that's when it'll come." After Tony does get his first call—a car with its engine on fire—he explains, "I got to spray water. I'm getting closer." In response, Hanlon narrates directly to the audience: "Listen, Tony was getting closer, but for the record, that was some flame, it wasn't a real fire." Having yet to experience a "real fire," the film suggests, Tony is still only a kid playing at a man's job.

All of this is made still clearer as the film and the firefighters it chronicles look back on the days leading up to September 11 and the death of a young firefighter in particular. As the scene opens, a newspaper headline fills the screen: "Rookie fireman dies at blaze." In an insightful statement, James Hanlon explains his recollections of the incident:

> I look back to last summer and it doesn't just seem like a different time. It seems like a different world. At the time we didn't think there could be anything worse than losing a single firefighter. We were all just kind of innocent—especially Tony.

Looking back in this manner, the film suggests the profound sense of innocence that, in retrospect, seems to have dominated the firehouse. For the entire firehouse, and for Tony in particular, the film suggests that September 11 will be a profound rite of passage.

Chaos and Fear: Masculine Bodies in Crisis

For the CBS broadcast of *9/11* (but not for the later released documentary), Robert De Niro offered a series of introductions to various parts of the film. Perhaps the most powerful of these frames the events of September 11 themselves as they are depicted within the documentary. After a commercial break showing images of male and female firefighters from around the country, De Niro narrates:

> Firemen live to help others' live. It's that simple. Everyday they wait for the call. Every day that passes without that alarm is a blessing and a burden. They know it's only a matter of time before they have to put their lives on the line. So they wait. What you are about to see is how brave men work under stress surrounded by chaos. They trained all of their lives for this

moment but nothing could have prepared them for what was about to happen.

The next scenes—no doubt the ones that most of the audience tuned into see—depict the events of September 11 from an insider's perspective. As Jules and Gedeon's cameras follow a group of firefighters within and around the World Trade Center, the film depicts their profound fear and suffering. Having established the strength and brotherhood of their firehouse, *9/11* now shows what happens when this strong brotherhood is threatened. In the process, the movie depicts an intense version of the sense of masculine crisis that dominated the middle to later 1990s.

The scene begins at 8:46 on the morning of September 11, with a group from the firehouse, including Jules Naudet and his camera, going to investigate an odor of gas in the street. In what has since become infamous footage, Jules turns his camera towards the World Trade Center to capture the first plane as it crashes. "Right then and there I knew that this was going to be the worst day of my life as a firefighter," one of the group members narrates as he remembers. "It's like the world just stopped," another recalls. Stressing the intense sense of chaos they felt at the moment, another firefighter remembers that "as we swung around in front of World Trade, my mind tells me, wow this is bad." "What do we do? What do we do for this?" still another firefighter remembers asking himself. The intense anxiety that this footage demonstrates stands in stark contrast to the collected, joking firehouse captured in the early part of the documentary as well as to the standard cultural conception of firefighters captured time and time again within American popular culture.

In narrating this story of chaos and fear, the film *9/11* tells several interrelated stories: the story of Jules and the group of firefighters within the World Trade Center itself; the story of Jules and Gedeon's concern for and quest to find each other; and the story of Tony as he figures out how to react to this first catastrophe. Inside the World Trade Center, Jules' camera and narration—along with carefully selected interview footage with various firefighters remembering the event—work to capture the intense fear of the firefighters who start their way up the stairs of the building. "The lobby looked like the plane hit the lobby," one of the firefighters remembers. "All of the damage was done already so you knew it was gonna be worse when we got upstairs." As Jules explains in another voiceover, "Companies come in. You see them with a concerned look on their face and they're sent up." Still, Jules and the others remember, they felt confident that the firefighters would get the situation under control. One of the firefighters explains, "I felt the mood that we we're gonna put the fire out. Everyone seemed confident. I knew I was," and Jules remembers thinking, "They'll put it out. That's what they do."

In this manner, the documentary reiterates a societal expectation at the same time that it shows that expectation beginning to crumble. The opening scenes of the film, discussed above, showed the strong, competent brotherhood of a stereotypical firehouse, establishing the mythic image of the firefighter that most Americans likely want to believe is true. These scenes within the World Trade Center, however, puncture much of those myths, showing these firemen as all too human, and in stark contrast to the typically macho depictions presented in movies such as *Backdraft* and in television programs such as *Emergency*. These firemen are not superhuman hard bodies, the movie suggests, but real men with their own fears and anxieties. Another of Jules' comments seems to make this clear. "I was seeing the look on the firefighters. It was not fear. It was disbelief. 'What's going on?' That made me panic a little bit. That made me panic."

Again, the interviews included within the film continually reinforce the fears that these firefighters experience as they enter the World Trade Center, further highlighting the firefighters' own sense of panic. As people begin to jump from the upper floors of the building, one of the firefighters remembers thinking, "How bad is it up there that the better option is to jump?" This feeling of fear becomes more pronounced after the first tower falls and the firefighters stop their mission to climb to the upper floors of the building and start simply trying to get themselves out. "For the first time, I looked in someone else's eyes and saw fear, which you don't see with firemen." Including a variety of images of ash-covered firefighters running to escape the building, the film works to capture the sense of chaos that the firefighters remember feeling. "I don't even know if I was touching the stairs on my way down," one firefighter recalls. "When I got to about [floor] three or two is when I started thinking of my family and I thought I gotta get out of here." Against an image of an ashen firefighter with a bloody nose, sitting down to rinse out his mouth with water, another firefighter comments, "Everyone's been wanting to ask me what happened, what happened. I said, Hell. Hell is what happened."

At the same time that the film depicts the struggles of these firefighters to leave the building, it also chronicles brothers Jules and Gedeon in their fear of having lost one another, making these two filmmakers main characters in their own movie. For instance, once Jules finds himself in the lobby of the World Trade Center, he remembers thinking that his brother was likely climbing the stairs with the rest of the firefighters. "As far as Jules knew," James Hanlon narrates, "Gedeon had followed Tony, the probie, into the tower." "For me," Jules explains on camera, "my brother was going up the stairs." In the meantime, Gedeon, who is actually still back at the firehouse, picks up his camera and begins to walk towards the World Trade Center. Says Hanlon, "He was sure his brother was inside, and he wanted to get to him."

In this manner, the film depicts the events of September 11 as a sort of struggle and rite of passage not only for these firefighters but for Jules and Gedeon as well. Addressing the camera directly and recalling his own reaction after the first tower fell, Gedeon remembers:

> I wondered for the first time if Jules is still alive. I never thought about it that way before. I realized that Jules could be dead.... I was feeling so responsible. I was the one that put him in this situation. I had to find Jules.

Similarly, after Jules leaves the World Trade Center with a fire chief and finds himself running down the street to escape a billowing cloud of smoke, Jules remembers asking himself, "Where is my brother? I start realizing I've probably lost my brother." He adds, "At that point, I think he's dead. It becomes too overwhelming." "The only thing I could think about was Jules," Gedeon explains once again, "and I remember thinking that if I survived, that I would be a better brother." Similar to the firefighters, the movie tries to suggest, Jules and Gedeon's own brotherhood plays an important role in their struggles to survive the events of September 11.

Finally, the movie also works to portray the probie Tony's responses to these events, framing this as the beginning of his masculine coming of age. Told to remain at the firehouse to answer the phone, Tony is alone in the building as the other firefighters go off to fight the fire. Gedeon, who returned to the firehouse after unsuccessfully attempting to make his way to the World Trade Center, comments on Tony's demeanor. Gedeon explains that Tony was "freaking out and swearing," suggesting that "Tony was expressing what we all felt." Seeing this as a moment of growth for Tony, Gedeon adds, "At that point I saw the fireman in him taking over." "Tony just wanted to go there," Gedeon says a moment or two later. When a retired fire chief arrives at the station, Tony joins him in going to the aid of the firefighters already in the towers—"the probie and the retired chief," James Hanlon emphasizes. "They're my firefighters," the retired chief says as he grabs his equipment in preparation to leave.

The combination of chaos and brotherhood illustrated within this documentary works to drive home the conflicted coming-of-age story the movie tells. *9/11* presents the chaos of September 11 as a powerful rite of passage for Tony, Gedeon, and Jules, as well as the firefighters who go inside the World Trade Center. Here, the movie presents these various figures as both eminently heroic and eminently human, empowered by their intense brotherhood at the same time that they are distraught by the intense horror of September 11. As one firefighter explains:

> You start to feel your anxiety build up and you take a deep breath and you say it's gonna be all right, let's keep going. Brothers ahead of me, brothers behind me, we're in this together. We're fighting together and were gonna do what we have to do.

In this way, the documentary served as an allegory for the American public more generally, presenting the audience with images of living human beings struggling to deal with the senselessness of September 11. It also performed the powerful mix of vulnerability and heroism that characterized the masculinity of the 1990s and continued at the turn of the century.

Intimacy and Togetherness: Unity in the Bonds of Brotherhood

In a review for the *New York Daily News*, television critic David Bianculli called *9/11* "humanity at its best," claiming "that in the midst of all that horrifying devastation is a tale of honor, duty, resilience, and love." Indeed, the last part of the documentary, which shows the return of the firefighters to the firehouse and the reunion of Jules and Gedeon, truly presents a triumphant tale, though it is a triumph mixed with pain and confusion. "It's very emotional," one of the fire-fighters remembers in another retrospective interview in front of the camera. "A lot of our guys are crying. I'm crying." As the camera shows images of members of the firehouse crying, hugging, and otherwise greeting each other, another firefighter comments, "It was a great thing to know that people are surviving this." Still another suggests, "It's not easy being a survivor. Why am I alive when so many others are dead?" "It was weird in a way," Hanlon explained, recalling something that happened after a day of digging for bodies in the rubble of the World Trade Center. "Walking back to the firehouse, people were cheering us, but we sure didn't feel like heroes." In these ways, *9/11* shows much that illustrates this same sense of tragedy-tinged heroism, offering a still more profound picture of the conflicted heroism that characterized the middle to later 1990s.

In a manner similar to the ideas of '90s masculinity guru Robert Bly, discussed above, *9/11* suggests that the suffering endured by these firefighters functions precisely to reclaim their heroic manhood. Sounding a bit like King Henry in Shakespeare's *Henry V*, the film makes it clear that it is precisely the brotherhood of these firefighters—this band of brothers—that gives them the strength to overcome the trials of September 11. After an emotional reunion with his brother Gedeon, Jules recalls the comments of the firefighters of the house. "Yesterday you had one brother," he quotes for the camera, "today you have fifty." "You just needed to be with the guys," another firefighter suggested, as he remembered the emotional aftermath of returning to the firehouse on September 11. Still another firefighter makes this more explicit: "You know, the only thing you have, the only thing that kept it all together was us, as a group, as a body, as a firehouse."

As might be expected, some of the most pronounced aspects of this story are told through the probie, Tony, whom the documentary had followed from the beginning. Discussing Tony, at one point Jules claims, "he proved himself that day to all of the guys." Likewise, James Hanlon explains towards the end of the movie:

> In the beginning they came to me and they said, let's make a documentary about a boy becoming a man during his nine-month probationary period. It turns out that Tony became a man in about nine hours, trying to help out on September 11

As far as Hanlon is concerned, this statement suggests Tony's willingness to pitch in and his experiences dealing with the tragedies of
September 11 have forced his development from a kid to a man.
Again, similarly to Bly's "Iron John," by digging within himself and
facing his fears, Tony seems to have embraced his masculinity and
come out as a new man.

While this might tend to make *9/11* sound like a fairly standard
coming-of-age story—the kind typically depicted in a variety of Hollywood war movies—the conflicted way in which the film narrates
this story gives it a different tenor. In this manner, *9/11* reflects the
troubled masculinity of the '90s more so than the give-em-hell rite-of-
passage stories of a number of other moments in time. As Tony recalls
his own experience on September 11, he remembers talking with a
firefighter who recounted the troubling things he had seen that day.
"'It was raining bodies,'" Tony recalls the man explaining, "and just
the way he said it, [you knew] that man had been through hell."
Speaking to the camera, but looking down at floor, Tony continues,
"There's gonna be a lot of pain to deal with in the future." Similarly,
the chief of the firehouse, Chief Pfeiffer, talks about the pain he feels
after having lost his brother—a firefighter for another New York City
firehouse—on September 11. "I was remembering how my brother
and I used to love being downtown, and doing this job, and how now
I didn't love it anymore."

Showing these firefighters as both eminently masculine—as with
Tony's coming of age—and eminently defeated, *9/11* illustrates much
of the same conflicted masculinity that characterized the popular culture of the '90s. Coming at the close of this decade, it makes sense that
the images of masculinity within *9/11* would resonate with the
American public. The 1990s were largely about creating images of
masculinity that upheld traditional masculine ideals while embracing
a sort of new masculinity that was more in touch with contemporary
critiques of this traditional manhood. As a result, the decade saw a
host of male heroes that were both strong and vulnerable. The heroes
of CBS's *9/11*, like so many other images given within the popular culture immediately after September 11, offered a still more pronounced
version of this same heroics.

The fact that this imagery could speak to the American public
makes sense in the context of September 11 itself. As many people
made clear at the time, the country felt shocked that such an attack
could happen within the United States, a country that had previously

seemed invulnerable. The attacks of September 11 offered a further extension of the feeling of crisis that many had identified with the 1990s, but now on a national level. The all-powerful, untouchable United States, which had prospered greatly during the 1990s, had been dealt a powerful blow by what would seem a less-than-worthy adversary using the least sophisticated weaponry imaginable. The images of CBS's *9/11*, like so many others, resonated with this sense of unimaginable vulnerability, while simultaneously suggesting that the country was still strong. Discussing the firefighters attempts to dig out the bodies buried under the rubble of the World Trade Center, one firefighter suggests that:

> There's something special, you know, when guys are relentless and just go-ing back and forth with nails in their hand. Taping it up. Gashes, blood eve-rywhere. Just taping it up and saying let's go back, let's see what we can do to make this situation a little better.

These conflicted images of vulnerability and strength once again used masculinity as a cultural resource, here using it make sense of this horrific tragedy. If even firefighters were suffering, than the blow had to have been profound; but the fact that they could fight back from such serious losses suggested that the entire country could.

Sensitivity vs. Cowardice: Masculinity and Citizenship in the Oughts

As a result of the losses of September 11, the United States earned a new victim status that would have felt particularly strange during the seeming glory days of the 1990s. William Saletan, a political commen-tator for online magazine *slate.com*, captured this clearly, if somewhat inadvertently, in a commentary written on September 18th. Here, Saletan responds to the "consequentalist" position that had attempted to connect the tragedy of September 11 to activities by the United States in the Middle East:

> Imagine yourself as a battered wife. Every so often, your husband gets an-gry and hits you. Why? You struggle to understand the connection between your behavior and his response. What are you doing that causes him to re-act this way? You hope that by identifying and avoiding the offending be-

haviour, you can regain domestic peace and a sense of control. You're de-
luding yourself. As long as your husband decides which of your acts will
earn you a beating, he's the master, and you're the slave. This is the problem
with the consequentialist argument for revising U.S. policy in the Middle
East.

By casting the United States as the battered wife to the battering husband of Middle East terrorists, Saletan performs a powerful sort of
cultural remembering. Whereas it would have made little sense to call
the United States the battered wife or slave of the Middle East during
the 1990s, such arguments can more easily enter the popular imagination in an era of vulnerability.

In framing September 11 as an instance of spousal abuse, Saletan
also makes this a highly gendered relationship. In this way, he illustrates another idea that characterized much of the masculinity crisis
of the '90s in that he uses a popular feminist position for less than
feminist aims. Indeed, it was only through the work of feminists that
spousal abuse eventually became recognized as a crime and it is
feminists who continue to advocate on behalf of abused wives. As the
injured, vulnerable country, Saletan suggests, the United States is like
a battered wife. What this ignores, of course, or rather hides, is the
profound power imbalance between the United States and the Middle
East (which is, indeed, a source of the sense of intense vulnerability
on the part of the United States—that it could be so clearly injured by
a so much weaker opponent). Regardless of how one views the mo-
tives of the September 11 attackers, it should be easy to see that the
United States holds economic and political power over the Middle
East, not the other way around.

Saletan's position, in a still more blatant manner than the documentary *9/11*, extends the sense of "sensitivity" that played a role in
the masculinity of the '90s. For Saletan, the United States is a hyper-
vulnerable battered wife being manipulated by the likes of the Middle
East, all the while trying her best to smooth things out. Still more so
than the images of '90s men discussed within the previous chapters,
Saletan's position takes new-age sensitivity to a hyperbolic extreme.
Similarly, just as the '90s men were profoundly schizophrenic, mixing
a new-age sensitivity with traditional ideals of masculine toughness
(as well as other traditional ideals about masculinity), Saletan's vision
is still more conflicted. As the vulnerable, well-meaning battered wife,
Saletan suggests, the United States is completely justified in retaliat-
ing against her abusive spouse. Ending his discussion with a quote

from Donald Rumsfeld, Saletan stresses the ultra-masculine toughness that becomes part and parcel of his image for turn-of-the-century
masculinity:

> The terrorists who struck the Pentagon and the World Trade Center "are
> clearly determined to try to force the United States of America and our val-
> ues to withdraw from the world," Defense Secretary Donald Rumsfeld ob-
> served yesterday. "We have a choice: either to change the way we live,
> which is unacceptable; or to change the way that they live. And we chose
> the latter." Amen.

In framing these attacks as a matter of spousal abuse, Saletan makes
the U.S. response into another case of battered woman syndrome, and
who can blame a woman for lashing out at her hyper-abusive husband?

Saletan's discussion, while extreme, captures the tenor of one
popular post–September 11 climate of masculinity. While on the surface this might not seem to have much to do with masculinity, the
conflicted sensitivity and toughness brought out in these discussions
carries over much of the images of manhood that characterize the
'90s. But while the '90s heroes' hypermasculine toughness stood as a
counterbalance to his new-age sensitivity, the hypermasculine toughness of the post–September 11 period is often seen as justified by the
intense vulnerability and pain experienced by the country as a result
of this tragedy. With images of wounded firefighters presented in
newscasts, magazines, and elsewhere, who could blame the U.S. government for seeking retribution? Likewise, whereas the conflicted
masculinity of the '90s was used to sell television programs, movies,
and presidential candidates, the stakes are much higher in these later
discussions, which seek to sell wars, regime changes, and a host of
other large-scale political actions.

Saletan's arguments illustrate one way in which ideas about gender and masculinity are subtly co-opted in order to justify particular
positions. After September 11, the U.S.'s conceptions of masculinity—
like so many other cultural understandings—became a battleground
for fighting out a whole host of understandings regarding these attacks. Just as Bill Clinton's sensitivity had been celebrated one minute
and then decried another, so this period saw sensitivity fought over
and struggled through. Immediately following the attacks of September 11, the French paper *Le Monde* proclaimed, "We are all Ameri-

cans." As a result, the mainstream popular culture saw France as a sensitive, empathetic friend, attempting to understand the nation's pain. Later, as France resisted the U.S. invasions of Afghanistan and then Iraq, it was a country of effeminate cowards. Even as many liberals contested the invasion of Afghanistan, others supported it as a way to liberate the seemingly oppressed women living within the country's borders (an idea the Bush administration quickly latched onto). Here, the meanings of sensitivity and gender sensitivity in particular were rewritten and fought over from one moment to the next.

These next two sections explore this conflicted sensitivity more fully. The first section looks more closely at the comments of Lance Morrow, illustrating his profoundly antisensitivity position and the rhetorical force of this argument. This section also explores the post–September 11 controversy surrounding Bill Maher's program *Politically Incorrect*. By questioning whether or not the attackers were truly cowards, Maher was himself framed as insensitive, demonstrating the strangely changing conceptions and uses of these various concepts. The second section explores several popular masculine heroes of the early 21st century, exploring the ways in which these heroes perform these same conflicted notions of sensitivity in extremely conflicted ways.

Morrow versus Maher

As the above examples illustrate, the aftermath of September 11 created an awkward and conflicted climate in terms of our mainstream cultural understandings of sensitivity. In many regards, people were expected to be more sensitive about the ways in which they talked about and depicted the country. Immediately after September 11, several Hollywood films were pulled, edited, or otherwise changed to make them more appropriate for American audiences. For instance, Ben Stiller's film *Zoolander* saw the twin towers digitally removed from one scene, for fear that their image might somehow be disrespectful to the memory of the deceased. Similarly, theatrical trailers for the upcoming film *Spiderman* were pulled because they featured an image of the twin towers with a spider web stretched between them. The producers of the film *Collateral Damage*, which features Arnold Schwarzenegger as a fireman who fights terrorists in order to avenge the death of his wife and children, decided to postpone the film's release, for fear that the subject matter would be too much for the American viewing public. Soon after September 11, debates ensued over how long college and professional football games would be postponed out of respect for the dead of the nation. "What you want to do is to make the appropriate response," National Hockey League Commissioner Gary Bettman said before canceling the league's preseason opening games. "You want to avoid making a decision that makes you appear insensitive." From sports commissioners to television and film producers, much of the nation was concerned over how best to be sensitive to people's reaction to this horrific tragedy.

This new climate of sensitivity also saw those who disagreed with the Bush administration's reactions to the event vilified in a variety of ways. The American Council of Trustees and Alumni, an organization sponsored by Lynn Cheney, among others, published *Defending Civilization: How Our Universities Are Failing America and What Can Be Done about It*. The report intended to call attention to university professors whose comments following September 11 seemed out of step with the Bush administration or national sentiment more generally. As the authors of the report explain:

> Even as many institutions enhanced security and many students exhibited American flags, professors across the country sponsored teach-ins that typically ranged from moral equivocation to explicit condemnations of America.... Some refused to make judgments. Many invoked tolerance and diversity as antidotes to evil. Some even pointed accusatory fingers, not at the terrorists, but at America itself.

The report goes on to list a variety of apparently offensive and otherwise insensitive comments, and faults professors with "intimidating" students into "conforming to a particular ideology," that the authors see as anti-American. Insensitive to the nation's suffering, as well as to their students' desires and fears, America's college and university professors failed to pay proper homage to the horror of September 11.

With athletic events postponed and college professors chastised for offending their students, a number of public voices called for a brand of self-censorship that would ensure a proper degree of empathy regarding the nation's loss. As a weekly columnist for the *Washington Post* explained in a January 4, 2002, editorial:

> John Ashcroft can relax because people have been listening to their Inner
> Ashcroft. I know this for a fact because I'm one of them. As a writer and edi-
> tor, I have been censoring myself and others quite a bit since Sept. 11. By
> "censoring" I mean deciding not to write or publish things for reasons other
> than my own judgment of their merits. What reasons? Sometimes it has
> been a sincere feeling that an ordinarily appropriate remark is inappropriate
> at this extraordinary moment. Sometimes it is genuine respect for readers
> who might feel that way even if I don't.

Trying to be respectful to the level of tragedy experienced on September 11, writers and others were to "watch what they say," to borrow the words of Bush's early press secretary Ari Fleischer. This extreme desire for sensitivity reflected much of the zeitgeist of the '90s. Someone risked being considered uncivilized, rude, or worse if they failed to pay sufficient respect to the public's sense of mourning and loss. Being extrasensitive was part and parcel of being a good person in the extraordinary climate of post–September 11 America.

An important incident involving Bill Maher's program *Politically Incorrect* captures the complexity of the cultural understandings of sensitivity and cowardice during this period. On an episode of the program aired shortly after September 11, Maher's guest, former Ronald Reagan aide Dinesh D'Souza suggested that the terrorists hadn't been cowards, because they had been willing to sacrifice their lives for their cause. This ran counter to the prevailing rhetoric of the time—from both George Bush and other politicians, and from the popular culture more generally—that insisted that the terrorists had committed a cowardly act in attacking the World Trade Center in the manner that they had. Maher agreed with D'Souza, and added, "We have been the cowards, lobbing cruise missiles from 2,000 miles away. That's cowardly. Staying in the airplane when it hits the building, say what you want about it, it's not cowardly." Maher's comments were seen as insensitive by a vocal portion of the general culture—led in part by Federal Express and Sears, who withdrew their advertising from the program following this incident.

A *Boston Herald* editorial parodied Maher's alleged defense of manly warfare. "Presumably, the loudmouth Hollywood lefty would have liked U.S. forces to show their manliness by going after al-Qaeda with bayonets or bare hands," the editorial's writer commented. "Maher's own idea of courage," the editorial continues, "is stacking his show with three guests who agree with him and one dissenter the gang proceeds to pummel. It's unlikely the Marines will use the comic's photo on recruiting posters." Even after Maher apparently tried to apologize for his comment, a writer for the *San Francisco Chronicle* had the following to say:

> Say what you want about political correctness and the danger of abridging freedom of expression in dangerous times, but what Maher said was ill-timed and stupid. And his rush to cover his backside with I'm-sorrys sounded less heartfelt and more like "Please don't take my show away and please don't pull your advertising and please don't make me a pariah for the rest of my career."

Seeing his comments as ill timed, his apologies as less than heartfelt, and his statements about courage as hypocritical, these journalists see Maher's remarks as generally insensitive. Apparently ABC agreed, as the network eventually agreed to remove the show from its lineup.

This incident demonstrates a strangely conflicted understanding of courage in line with the more generally conflicted understanding of sensitivity that characterized the post–September 11 climate in which these remarks took place. To suggest that the World Trade Center attackers were not cowards because they were willing to die for their cause—as Maher did—was seen as insensitive to the feelings of the public, if not just plain stupid. In trying to empathize with the terrorists—even in a fairly sarcastic sense—Maher was deemed insensitive to the American public. By exercising one form of sensitivity, Maher was neglecting another. Americans were supposed to recognize that the terrorists actually lacked courage; to suggest that these attackers were courageous, or somehow tough because of the mission they had been willing to complete, was to commit a grave cultural wrong.

But this emphasis on sensitivity was accompanied by an equally extreme call for toughness and violence. An editorial by Kathleen Parker in the September 26, 2001, edition of the *Milwaukee Journal Sentinel* illustrates this widely expressed sentiment:

> War demands much of a nation's citizenry, but only in America does war demand sensitivity training. We're so earnest, so caring, so apologetic as we go about the business of war, our enemies must smile in their sleep.

According to Parker, the country needs to toughen up, not only on the terrorists but within the United States as well. "President Bush was the first to get his wrist slapped for using terminology that some found offensive when he suggested that our war against terrorism was a 'crusade,'" Parker suggests. "We who are, presumably, insensitive knew just what he meant. Uncapitalized, the word means 'a remedial enterprise undertaken with zeal and enthusiasm,' according to my dictionary." Likewise, Parker argues that those who might be upset about racial profiling and other domestic responses by the Bush administration need to toughen their skin. For Parker, being part of the country means being able to put up with various "inconveniences." Put aside sensitivity, Parker urges, both at home and abroad.

Another response from *Time* commentator Lance Morrow further demonstrates the post–September 11 assault against sensitivity (Morrow, September 14, 2001). Here, Morrow calls for the American public to put aside their sensitivity and to get in touch with a good old-fashioned understanding of anger. "For once let's not have grief counselors standing by with banal consolations," Morrow begins his discussion, speaking explicitly against a late 20th-century idea about how to deal with tragedy. "For once, let's have no fatuous rhetoric about 'healing,'" Morrow continues. "Healing is inappropriate now, and dangerous. There will be time later for the tears of sorrow." In contrast to what Morrow sees as an overly new age understanding of sorrow, Morrow calls for a tougher response. "A day cannot live in infamy without the nourishment of rage," he continues further. "Let's have rage."

Again, in stark contrast to the hypersensitive climate demonstrated by some of the examples above, Morrow calls for an equally hyperbolic sense of anger and emotional toughness.

> What's needed is a unified, unifying, Pearl Harbor sort of purple American fury—a ruthless indignation that doesn't leak away in a week or two, wandering off into Prozac-induced forgetfulness or into the next media sensation (O. J. … Elián … Chandra …) or into a corruptly thoughtful relativism (as has happened in the recent past, when, for example, you might hear someone say, "Terrible what he did, of course, but, you know, the Unabomber does have a point, doesn't he, about modern technology?").

Recognizing that this might not be easy, Morrow observes, "A policy of focused brutality does not come easily to a self-conscious, self-indulgent, contradictory, diverse, humane nation with a short attention span." However, Morrow argues, "America needs to relearn a lost discipline, self-confident relentlessness—and to relearn why human nature has equipped us all with a weapon (abhorred in decent peacetime societies) called hatred."

As a civilized country, Morrow suggests, the United States has lost its sense of hatred and rage. However, Morrow thinks that the horrors of September 11 may be just what the country needs to get in touch with what it has lost in the midst of its 21st-century tolerance. "Is the medicine too strong?" he asks. "Call it, rather, a wholesome and intelligent enmity—the sort that impels even such a prosperous, messily tolerant organism as America to act." To those who might try to empathize with the attackers, Morrow offers an explicit message. "Anyone who does not loathe the people who did these things, and the people who cheer them on, is too philosophical for decent company." The United States, Morrow holds, needs to put aside the sense of sensitivity the country may feel as a result of its overcivilized nature. As Morrow puts it, "let the civilized toughen up, and let the uncivilized take their chances in the game they started."

Morrow's position, like that of Kathleen Parker and a host of others who made similar arguments shortly after September 11, urge the American public to toughen up. Alongside the equally prolific calls for a kind of new-age sensitivity about what one says and does (as illustrated in the example of Bill Maher and others above), the post–September 11 United States saw a bizarrely conflicted understanding of sensitivity. This combination of calls for sensitivity and cries against becoming weak through caring too much offers an interesting restatement of the masculine ideals about sensitivity and toughness that characterized the '90s. As careful and as caring as someone needed to be in terms of choosing their words for the public, they needed to be equally willing to use force and be tough, lest they be taken over by their enemies. What had been a general model for selling masculinity in the '90s had become a measure of communication ethics and social and political action more generally, though in a still more exaggerated form.

Likewise, just as the '90s crisis of masculinity and its accompanying ideas could be used in multiple ways and to make multiple arguments, so it is with the post–September 11 deployment of these similar ideas. So, for instance, the terrorists were cowards, even if they were willing to kill themselves in pursuit of a cause (in contrast, a main hero of the '90s film *Independence Day* is a father and retired Air Force soldier who pilots his jet on a suicide mission to blow up an alien ship). If someone empathized with the terrorists, he or she could be accused of being too sensitive him or herself (as were some of the "demurring" touchy-feely, new-age professors discussed in the *Defending of Civilization* report) or of being insensitive to the American public (as was Bill Maher). If someone were unwilling to sufficiently respond to Afghanistan, he or she could be accused of being cowardly, weak, or at least "too philosophical for decent company."

The multiple uses of these ideas also demonstrate how our conceptions of masculinity and gender more generally can subtly work to disguise the logic of various arguments. Sensitivity and toughness are both highly gendered terms, as I've suggested throughout the previous chapters. In Americanese, asking someone to "toughen up" is nearly synonymous with telling someone to "be a man." Similarly, courage is the property of firefighters and other American heroes; cowardice is the stuff of terrorists and of the weak-minded who cannot get behind the United States. When Bill Maher suggested that the World Trade Center attackers were not cowards, he essentially suggested that they had a claim to their masculinity—that they had died, like men, for their cause. In a country that generally sees masculinity and heroism as synonyms, Maher's comments were unthinkable and, according to his opponents, insensitive. While it might seem contradictory to celebrate hypersensitivity and brutal toughness at the same time, this makes sense in a climate that sees masculinity as both heroic and destructive, as simultaneously impenetrable and vulnerable, and as both powerful and disempowered. The masculine heroes of the '90s—from Bill Clinton to Steven Seagal and Leonardo DiCaprio—used these conflicted anxieties as a way to win votes or sell movies. At the turn of this century, these work to promote wars and silence people attempting to make particular kinds of arguments.

There is nothing intrinsically wrong with either sensitivity or toughness. At times we could all no doubt be tougher and at others we could stand to be more sensitive. However, as these phrases get used, along with phrases such as courage or cowardice, they too often allow people to make arguments without really having to support them. In a culture obsessed with masculinity, it's too easy to accuse someone of being a coward, just as it's easy to charge someone with being insensitive, particularly when we never stop to question these ideas. Perhaps Maher's comment about the cowardice of the WTC attackers illustrated this too clearly for the people who opposed him. If the terrorists themselves could be courageous, then what does courage really mean? When we fail to think critically about these terms, we risk falling into a logic that escapes us, caught in our own invisible acquiescence to masculine ideals and victim to the rhetoric of advertisers, propagandists, and ideologues who hope to capitalize on this cultural blindness.

The New Hypersensitive Killers: Vic Mackey and Tony Soprano

Just as the 1990s saw the creation of a particular brand of masculine hero, so has the post-Clinton era seen its own version of manhood. Far from an all-out backlash or rejection of Clinton-era ideals, however, this recent masculinity seems simply a more pronounced depiction of the same conflicted vision offered up in the previous decade, as I've already suggested above. George W. Bush himself, with his self-proclaimed "compassionate conservatism," illustrates this quite clearly. Of his autobiography, *A Charge to Keep*, published in 1999, Bush said that he "wanted to show people my heart," trying to call attention to his sensitivity and inner being. In February 2000, well into Bush's campaign for president, a *New York Times* reporter argued that "party elders beamed, because here, at long last, was a vibrant, youthful Republican candidate who could court voters across a wide range of the political spectrum and match any Democrat in the sensitivity wars." The Bush campaign fought hard to show that Bush was, indeed, a compassionate, sensitive candidate for president.

Likewise, from the early stages of Bush's campaign for the presidency, the public regularly commented on the startled, "deer-stuck-in-headlights" look that seemed to cross his face whenever he spoke in front of an audience—suggesting a still more hyperbolic vision of sensitivity. In December 1999, a writer for *Newsweek* sarcastically captured one of the Bush campaign's attempts to make him seem more at ease:

> As they prepped George W. Bush for debate, his handlers considered every detail. For example, should he wear a watch? Dad had worn one in his debate with Bill Clinton in 1992. The result was disaster. TV cameras caught the president glancing at the time, as if he wanted to leave the stage—or the White House. So maybe W shouldn't wear one. But he always does. Someone might notice its absence—and bring up 1992. Final answer: wear it, but, whatever you do, don't look at it. Nothing else was left to chance.

Framing Bush as a daddy's boy who needs his handlers to ensure he doesn't make a gaff, this writer captured the sense of humor surrounding Bush's apparent stage fright. Similarly, the large list of "Bushisms" that began to pile up—the verbal mistakes Bush seemed apt to make when speaking in public—further reflected his discomfort with speaking in front of audiences. "I know how hard it is to put food on your family," he said to a group of struggling workers. Bush was a terrified speaker whose tongue tended to wag more quickly than his brain. Likewise, the Bush administration moved quickly to defend his sense of courage when various commentators suggested that Bush had used Air Force One to flee the scene of September 11 immediately after the attacks. Whether in front of an audience or faced with a national disaster, critics suggested, Bush was an inexperienced child playing at an adult's job.

At the same time, Bush seemed the epitome of hypermasculine Texas culture. A writer for the *San Diego Union-Tribune* captured this in a discussion of this Texas boy turned president. "In walked George W." she explains of a younger Bush, "5 feet and 11 inches of swagger striding across the floor, his feet in cowboy boots, his cheek stuffed with a plug of tobacco, his blue eyes a glint of something tart and testy." Likewise, Bush's record on crime and capital punishment seemed to point towards an inherent toughness and hypermasculinity. In an October presidential debate moderated by PBS's Jim Lehrer, one participant made direct reference to this record in a question he put to Bush:

> In one of the last debates held, the subject of capital punishment came up. And in your response to the question you seemed to overly enjoy, as a matter of fact, proud that Texas led the nation in execution of prisoners. Sir, did I misread your response and are you really, really proud of the fact that Texas is number one in executions?

While Bush quickly reassured the audience that he was not proud of his record, he still maintained the sense of tenacity and toughness that such a record presumably suggested. In his response, he commented:

> One of the things that happens when you're a governor, often times you have to make tough decisions. And you can't let public persuasion sway you because the job's to enforce the law. And that's what I did, sir. There've been some tough cases come across my desk. Some of the hardest moments since I've been the governor of the state of Texas is to deal with those cases.

Similarly, when a group of Canadian citizens protested the execution of a man from Alberta, Bush remained resolute in his response. "We believe in swift and sure punishment," Bush commented at a news conference. "If you are a Canadian and come to our state, don't murder anyone."

As a president characterized both by intense stage fright and absolute resolve in putting criminals to death, George Bush offered an extreme version of the conflicted machismo depicted in the 1990s, making him well suited to early 21st-century American manhood. Similarly, his willingness to bare all of his personal history of drug and alcohol abuse and other youthful indiscretions made his masculinity all the more conflicted. While the act of confession had come to characterize a particularly '90s view of sensitivity—illustrated, in fact, in some of Clinton's own apologies—the fact that Bush had such a sordid history to confess also stood as evidence of his background as a rebellious good-old-boy. As framed by his campaign, Bush was a conflicted character who had repented his past but remained tough in the face of public pressure. Bush's version of masculinity further illustrates a larger, post-Clinton manhood that came to prominence with Bush's presidency as well as with the events surrounding September 11. Again, far from rejecting its '90s predecessor, this conception of manhood presents similar anxieties and tensions but in a still more schizophrenic form.

As with the '90s, the fictional imagery of the early 20th century reflects this same conflicted vision of masculinity. Two characters in particular, Tony Soprano of HBO's program *The Sopranos*, and Vic Mackey of FX's popular *The Shield*, depict the extremely contradictory masculinity of the turn of the 21st century. While one character, Tony Soprano, works as a mob boss, and the other, Vic Mackey, works as a police officer, these two figures nonetheless have much in common. For all of his mob dealings, Tony Soprano comes off as a fairly conventional father to his two children. Likewise, Vic Mackey manages to embezzle money and murder a fellow police officer, all the while making arrests and keeping the streets safe for his own family. Both Tony and Vic live in the netherworld between right and wrong, performing an act as a compassionate father or law-abiding citizen one minute, and then putting a gun to someone's head the next.

Tony and Vic demonstrate very similar brands of masculinity as well. In contrast to the hard-bodied male of the '80s or even the mostly fit sensitive '90s guy, Tony and Vic are both overweight, bald fathers. In this respect, they resemble Homer Simpson and Al Bundy more than most of the male heroes from the previous two decades (with a few notable exceptions, such as *NYPD Blue* character Andy Sipowicz, played by Dennis Franz—who portrayed a similar character on *Hill Street Blues* in the '80s). In contrast to Homer and Al, however, both Vic and Tony are take-charge, no-nonsense tough guys who usually get their way in their various business dealings. They have no compunction against using their muscle to intimate someone into going along with their positions. As tough as they both are, however, both Vic and Tony deal with a variety of psychic demons—Tony working out his various issues with his psychiatrist Dr. Melfi and Vic struggling to keep his marriage and family in one piece. Like the men of the '90s, Vic and Tony offer a thoroughly conflicted masculinity. However, in line with the early 21st-century climate I've already laid out above, theirs is a still more pronounced schism. Because Tony and Vic are both chillingly ruthless, even slight acts of kindness or vulnerability prove a stark contrast.

Tony Soprano: The Sad Clown

In a *Newsweek* article discussing the popularity of *The Sopranos*, writers Mark Peyser and Yahling Ching make reference to the conflicted nature of Tony's character. On *The Sopranos*, they recognize, "even a murderous thug like Tony has to go to a shrink and pop Prozac." Indeed, the conflicted nature of Tony's character—especially relative to mobster characters of the past—plays a central role in the program from its opening episode onward. The series begins in the lobby of psychiatrist Dr. Melfi as Tony awaits his first counseling session. Having recently experienced a massive panic attack, mob boss Tony must now share his feelings with Dr. Melfi in what would become the most provocative story line of the program. "Look, it's impossible for me to talk to a psychiatrist," he tells Dr. Melfi at the beginning of this first session, as this comment is clearly meant to be taken in at least two ways. One the one hand, he can't talk to a psychiatrist because it might mean exposing his own work in organized crime and, by implication, that of his "family" as well (if anyone finds out that he is talking to Dr. Melfi, he later tells his wife, Tony will likely receive a "steel jacketed antidepressant in the back of the neck"). On the other hand, Tony can't talk to a psychiatrist because it suggests a kind of intimacy and vulnerability that seems antithetical to his tough-guy attitude and persona. The fact that Dr. Melfi is a female doctor seems to make this all the more pronounced. Sharing his innermost feelings with a female psychiatrist runs against all of Tony's stereotypical hypermasculine attributes.

The stories that Tony relates during this first session further highlight the conflict between various parts of his character. For instance, at the beginning of the session, Tony recounts a story about a group of ducks who made a home in his pool—setting up another story that the series would come back to at various times in future episodes. "A couple of months before all of this," Tony begins, "these two wild ducks landed in my pool." As Tony tells this story, the scene cuts from Dr. Melfi's office to Tony's backyard, showing him in his bathrobe as he looks for these ducks a few days earlier. "It was amazing," he continues, clearly touched and excited by this story. "They're from Canada or someplace and it was mating season. They had some ducklings." As the scene continues, Tony climbs into his swimming pool, bread in hand, to feed these ducks and ducklings. "If you don't like the ramp, I'll build you another one," he says to them. Next, as the ducklings flap their wings, trying to fly, Tony excitedly calls to his kids to come and watch. Clearly enamored with these ducks and their ducklings, this scene helps to establish the sensitive, loving aspects of Tony's character.

In contrast, the next story that Tony recounts, which follows immediately after this warm story of duck family life, helps to emphasize Tony's cold, hyperbolically tough nature. Driving to work with his nephew Christopher, Tony spots a man who owes him money that he has been unable to repay. As Christopher stops the car to try to confront him, the man runs away, heading through the lawn in front of a larger office building. As Christopher pursues the man on foot, Tony slides into the driver's seat of Christopher's car and begins chasing the man himself, driving the car over sidewalks and through the grass lawn, people in business attire jumping to get out of his way. Highlighting the pleasure Tony seems to take in this sort of chase, the camera captures his smiling face as he turns the car onto the sidewalk. Likewise, the upbeat song "I Wonder Why" by Dion and the Belmonts accompanies the chase, suggesting that this is a moment of pleasure rather than fear. In contrast to the sensitive duck-loving character in the earlier scene, this is a bloodthirsty killer who takes pleasure in others' pain.

When Tony catches up to the man, he hits him with the car, sending him sideways into the grass. As the man whimpers and cries, Tony gets out of the car and approaches him, taking his own rings off his fingers and placing them in his jacket pocket, preparing for the punches this scene suggests Tony feels all too familiar delivering. "Are you all right?" Tony asks the man with feigned concern. "My leg is broken," the man cries from the grass, "the bone's coming out." "Let me see," Tony replies, again feigning concern. Next, Tony punches the man in his broken leg, adding, "I'll give you a fucking bone, you prick. Where's my fucking money?" Tony then punches the man repeatedly in the face, shouting out a variety of profanities as he does so. A crowd of onlookers stands by, obviously aghast that someone could commit so vicious an act in broad daylight. Calling Christopher over to continue the beating, Tony seems to care little about the crowd watching him. "Shut up!" he says to the crying man, leaving him beaten and broken in the grass. "You prick," he adds again as they get in the car to leave. Maliciously injuring this man, Tony here demonstrates his profound hypermasculine toughness—a stark contrast from the earlier scene with the ducks.

The next scenes, which also illustrate stories Tony shares with Dr. Melfi, seem to highlight Tony's sensitivity once again. Talking about his Uncle Junior, for instance, another top person in his mob family, Tony laments the conflicted relationship they have with one another. "I love my uncle," Tony explains to Dr. Melfi. "At the same time, when I was young he told my girl cousins that I would never be a varsity athlete, and, frankly, that was a tremendous blow to my self-esteem." Making such comments to Dr. Melfi, Tony seems to highlight his own conflicted, vulnerable character, working against the stereotypical image of a toughened mob boss. Similarly, the next scenes show Tony as he attempts to be a loving son to his mother, Livia. Going to her house, Tony brings a new CD player and several CDs, trying to get his mother to dance with him in her kitchen and kissing her playfully on the cheek. As she resists, Tony explains, "Ma, you need something to occupy your mind. When Dad died you were gonna do all kinds of things." Trying to play the role of dutiful son, even as his mother pushes him away (indeed, in a later episode she tries to have him killed), Tony seems to have a sensitive loving heart, despite what his violent, coldhearted actions might suggest.

Continuing his conversation with Dr. Melfi, Tony explains that his panic attack occurred just after the ducks and ducklings in his backyard flew away, a further suggestion of his intense connection to them. However, as Dr. Melfi tries to get him to open up about this, Tony resists, taking refuge in his tough-guy persona once again.

> Let me tell you something. Nowadays everybody's got to go to shrinks and counselors and go on Sally Jesse Raphael and talk about their problems. What ever happened to Gary Cooper, the strong, silent type? That was an American. He wasn't in touch with his feelings. He just did what he had to do. See what they didn't know is that once they got Gary Cooper in touch with his feelings that they wouldn't be able to shut him up. Then it's dysfunction this and dysfunction that.

Using Gary Cooper as a metaphor (and still another figure that he will bring up in future episodes of the program), Tony here laments the disappearance of a more stereotypically traditional masculinity. Though in this, Tony clearly laments his own masculinity as well. Forced to share his problems with a psychiatrist, rather than handle them silently by himself, Tony seems out of touch with the version of masculinity illustrated within his more toughened character shown at other moments throughout the series.

"Do you have any qualms about how you actually make a living?" Dr. Melfi asks Tony at a second session, after Tony has experienced yet another panic attack. "Yeah," Tony responds, "I find I have to be the sad clown, laughing on the outside, crying on the inside." Again lamenting the fall of a more traditional masculinity, Tony explains:

> See, things are trending downward. Used to be, a guy got pinched, he took his jolt in prison no matter what. Everybody upheld the code of silence. Nowadays, no values. Guys today have no room for the penal experience.

Similarly, a few scenes later, Tony takes his daughter, Meadow, into an old church and sits down with her in one of the pews. "Your great-grandfather and his brother Frank built this place," Tony explains, celebrating a masculinity of the past. "Stone and marble workers that came over here from Italy and built this place." "They didn't design it, but they knew how to build it," Tony continues, as the camera pans the friezes lining the walls. "Go out now and find two guys who can put decent grout around the bathtub." Nostalgic for a mythic masculinity of the past, Tony sees himself, and most of his fellow men, as pale replicas of this more heroically manly past.

The episode's last scene in Dr. Melfi's office makes this theme still more explicit, making clear that Tony fears the disappearance of his own manhood. "I had a dream last night," Tony explains to Dr. Melfi. "My belly button was a Phillips-head screw and I'm working on screwing it. And when I get it unscrewed, my penis falls off." Continuing with his story, Tony narrates, "I'm holding it and I'm running around looking for the guy who used to work on my Lincoln when I used to drive Lincolns. And I'm holding it up in this hand," he explains, raising his hand, "and this bird swoops down and grabs it in its beak and flies off with it." Noting what she sees as an important connection, Dr. Melfi eventually turns the conversation towards the ducks that had landed in Tony's pool. "Those goddamned ducks," Tony exclaims, getting teary-eyed. "What was it about those ducks that meant so much to you?" Dr. Melfi asks him. "I don't know. It was just a trip having those wild creatures come to my pool, have their little babies. I was sad to see them go." Continuing to get emotional, Tony laments, "Oh Jesus Fuck, now he's gonna cry." As he begins weeping, Dr. Melfi passes him a box of tissues.

This scene serves to highlight the hyperemotional nature of Tony's character. It also illustrates the profound similarities and differences between Tony Soprano and the men of the '90s discussed in the previous chapters. Like '90s heroes Bill Clinton and Leonard DiCaprio, Tony Soprano offers a masculine figure simultaneously sensitive and tough. However, like the masculine zeitgeist of the early 21st century more generally, Tony's is a still more profoundly conflicted manhood. Leo's Jack Dawson illustrated his toughness through an occasional fistfight aboard the *Titanic* and his sensitivity through an affinity for art. In contrast, Tony's character moves from a vicious beating or killing to a crying fest in Dr. Melfi's office. Likewise, Tony's references to Gary Cooper, the penal experience, and the Italian stone and marble workers of the past, suggest his own awareness and fear that contemporary masculinity seems to be changing. Whereas Leo seemed relatively comfortable with his conflicted manhood, Tony seems tormented by his, making his struggle all the more profound. Reduced to sharing his feelings with a psychiatrist, Tony's character seems to see himself as a castrated ghost of some mythic masculinity past. A toughened killer who cries in his psychiatrist's office, Tony embodies the vision of masculinity that has him so upset in the first place.

Vic Mackey: Andy Sipowicz on Steroids

Michael Chicklis' character Vic Mackey in *The Shield* illustrates a similarly conflicted vision of masculinity.[2] A review of the program in the *San Diego Union-Tribune* described a like conflict within actor Chicklis himself, explaining that the series' creator had been "surprised when Michael Chiklis, until then a pudgy actor who specialized in nice-guy roles, showed up hard-bodied, bullet-headed and belligerent to audition for the role of renegade Det. Vic Mackey." An editor for the *Pittsburgh-Post Gazette* made a similar point when the program began, telling audiences to "forget the Michael Chiklis you know—the round, puffy family man who presided over 'The Commish' and 'Daddio.' As bad cop Vic Mackey on FX's new cop drama 'The Shield,' Chiklis has become a bald tough guy." Calling attention to Chiklis' past performances, these writers ultimately stress the central conflict within Vic Mackey as well. As a writer for the *Houston Chronicle* puts it, Vic is "the guy you call when you need a confession beaten out of some lowlife, or when you want to send a message to a punk drug dealer." However, he adds, "He's not heartless. He's a family guy." Like Tony Soprano, Vic Mackey maintains a vision of 21st-century manhood that seeks to highlight both hypertoughness and hypersensitivity.

"Good cop and bad cop have left for the day," Vic tells a kidnapping and rape suspect in an early episode as he prepares to "interrogate" him; "I'm a different kind of cop." Indeed, Vic Mackey's character offers a vision of police work far from the squeaky-clean likes of Joe Friday on the television series *Dragnet*. Rather than trying to coax or threaten this suspect into confessing, Vic simply pulls out a telephone book and strikes him repeatedly until he tells him where he has hidden the young victim. Such actions might not seem that new for a movie or television police officer; after all, rogue police officers are a common feature in much media fare. *Hill Street Blues* had its share (for instance, undercover officer Michael Belker, played by

Bruce Weitz), as have movies such as the *Lethal Weapon* series that depict officers who pay little attention to procedures in making their various arrests. In that Vic often goes his own way in solving the various crimes he investigates, he carries on a long tradition of badass cops who seemingly cut through the bureaucracy of the police department in order to keep the streets safe.

However, Vic Mackey seems to take this to a new level. In the first episode of the program, he and his team break into the house of a drug trafficker named Two-Time. After shooting and killing Two-Time (who shoots at Vic and his team first), Vic takes the gun from his dying hand and turns and shoots another member of his own team in the head. Knowing that this newest team member has been secretly working with their police captain—who himself hopes to prove that Vic and his teammates are stealing drugs and embezzling money—Vic kills him rather than risk getting himself and his team into trouble. In a later episode, Vic's team hijacks a shipment of Armenian drug money, hoping to keep that money for themselves. In yet another episode, Vic and his teammates illegally hold a professional basketball player in a hotel room, hoping to help the Lakers in that evening's game. Far from the traditional television police officer, Vic is the dirty cop made hero.

In this manner, like Tony Soprano—whose calls for Gary Cooper lament the fall of a more traditional manhood—Vic seems to offer his own a critique of the sensitive masculinity of the previous decade. In one conversation, the officer whom Vic will eventually kill says to him, "My dad was a cop. He admired guys like you." "Ah, well," Mackey responds, sounding like Tony Soprano remembering the stonecutters of the past, "those were some honest to goodness men back then, huh?" Similarly, Vic's character seems to offer a critique of the supposed political correctness of the '90s. For instance, Vic clearly has no use for the cultural sensitivity that many called for during the previous decade and that seemed a component of many '90s men. In a conversation about David Aceveda, the Latino man recently appointed captain of the department in which Mackey works, Vic comments "What do you expect from a damn quota baby?" Mackey makes similar comments in a conversation with Captain Aceveda in front of a number of other officers. "In this building, I'm in charge," Captain Aceveda shouts at Mackey across the police station. "Well, maybe in your own mind, amigo," Mackey jeers back, adding, "But in the real world I don't answer to you. Not today, not tomorrow, not even on Cinco de Mayo." Using his best overly exaggerated Spanish accent, Vic seems to make the racism of his comment as obvious as possible.

With this hypertough, nonpolitically correct persona, Vic Mackey seems a far cry from the sensitive new men of the '90s. Here, the show works to suggest, is a hard-core man's man. However, just as with Tony Soprano, Mackey's character demonstrates some odd and interesting conflicts as well. One such story line involves another police officer, Julien, who turns out to be a closeted homosexual, trying to fight his own homosexual tendencies and make himself straight. Ashamed of his own sexual activities, Julien tries to keep his sexuality a secret from the rest of the police department. After Julien witnesses someone on Mackey's team stealing two bricks of cocaine from a crime scene, Vic busts in on Julien and his boyfriend, Tomas, at the boyfriend's house, ostensibly to serve an open warrant for the boyfriend's arrest. "Your boyfriend will come visit you in lock up," Vic says to Tomas, "Won't you, officer."

In catching Julien and Tomas together, Mackey plans to blackmail Julien into retracting a statement he has made to Captain Aceveda. Here, Vic's anti-politically correct character stands out still more clearly. Willing to use Julien's sexuality against him, Vic rejects the sensitivity of multiculturalism. "I pulled some strings and got you released until your court date," Vic explains to Tomas in front of Julien the next morning. "I just need some help with one thing first," he continues. "In filling out my arrest report I'm a little confused. When I busted in it looked like you had some cop's dick in your mouth. That can't be right, can it?" Later, Vic confronts Julien in the bathroom of the police station. "I'm willing to keep your secret," he tells Julien; "What do you think I want to twist you arm?" As Julien refuses to retract his statement, Mackey explains, "Handing in this arrest report is the only way I can think to stop you from making a huge mistake." "That report doesn't prove anything," Julien insists. Vic laughs at Julien's comment, well aware of the strong homophobia of the police station. "I don't have to prove you're gay," he comments. "In this house, all I gotta do is say it—with all the gory details." Scared of having Vic expose his sexuality to his fellow officers, Julien agrees to retract his statement.

While this incident seems to clearly highlight Vic's rejection of cultural sensitivity, his subsequent relationship with Julien seems to frame this in still different ways. In a conversation after Julien has retracted his statement, Julien tries to clarify himself to Vic. "I'm not gay," he explains. "Julien, come on," Vic responds. "I'm not," Julien continues, "it's this thing inside of me. I push it down, it goes away, but then it comes back stronger." Nearly in tears, Julien adds, "I am so weak. I hate this thing inside of me." In a more caring, sensitive voice, Vic responds, "Julien, you can't go through life hating who you are." Here, Vic seems more informed and caring than Julien himself, as well as Julien's reverend, who later tries to convince Julien that he simply requires prayer and "sexual reorientation." Likewise, in subsequent episodes, other members of the police station will find out about Julien's sexuality and begin harassing him. Vic, the hypermasculine tough guy, seems unfazed and unconcerned. "I've pushed away everyone.... I'm alone," Julien tells Vic at the end of this conversation. "Hey, no you're not," Vic responds. "You were there for me. I'm there for you." Ever conflicted, Vic seems to have no problem with the fact that Julien is gay; yet he also has no problem with using Julien's sexuality to blackmail him into silence.

Vic's relationship with Connie, a prostitute who also serves as one of Vic's informants, further highlights his conflicted character. Emphasizing his sensitivity, Vic's various interactions with Connie offer glimpses of his capacity for compassion and concern. In one episode, Vic makes a vow to help Connie overcome her addiction to crack cocaine. After leaving Connie with a friend of his in a hotel room, Vic takes her infant son, Brian, to stay with his wife and family. "Why don't you just take the crack whore to detox?" Vic's wife insists when Vic shows up with Brian. "I can't," Vic explains, "she'll lose him if I do." At his wife's request, Vic explains his connection to Connie and his devotion to her and her son:

> I pulled her out of a brawl two years ago.... I gave her my number. I told her to call me if she ever needed help and one night she did. Cracked-out on the floor of a hotel bathroom, six months pregnant, screaming, crying, begging for forgiveness. I show up and find her in a pool of bloody crystals. She tried to get rid of him with some Drano and a plunger handle. I rushed them to Mission Cross. They did an emergency C-section.... He's a tough kid.

Affected by this early experience with Connie and Brian, Vic feels a connection to both of them and does his best to help them however he can. As tough as he may be, he has a soft spot for this mother and child.

Another incident with Connie helps demonstrate Vic's devotion to her, while also illustrating his conflicted mix of sensitivity and toughness. In this scene, Connie calls Vic from a hotel room after shooting one of her clients she mistakes for a serial killer in a crack-induced hallucination. "What did you do here?" Vic asks Connie in disbelief. As Vic prepares to call the police department and report the shooting, Connie insists, "All I have is Brian. Don't let me lose him." Thinking, Vic instructs Connie to tell the police that the customer had refused to pay then began beating her, forcing Connie to shoot him in self-defense. "Your friends, they'll believe that?" Connie asks Vic. "Only if you have the bruises to prove it," Vic responds. "Okay," Connie answers, closing her eyes as Vic makes a fist. Teary eyed, Vic draws back his fist. "No, I can't, Connie." Vic stops himself, unable to strike her. "It's okay, Vic. C'mon," Connie adds again. With Vic repeating the word no, Connie screams at him to hit her, slapping him in the face. When Vic still refuses, Connie takes old of his hands tearfully begging, "Please, Vic, please." Calming himself, Vic holds Connie's head between his hands and kisses her gently on the forehead. "Hit me," Connie whispers to him again. His eyes filling with tears, Vic pulls back and strikes her several times to the face.

Vic's difficulty to hit her serves to highlight the conflicted nature of his character. Like Tony Soprano, who can blithely kill someone one minute and in the next find himself crying in Dr. Melfi's office, so Vic seems an extremely troubled tough guy. As tough as he may be, this scene seems to suggest, striking Connie is just too much for him to stomach. Although Vic himself concocts this plan, only Connie's repeated requests can bring him to carry it out. In short, while Vic seems to be a sadist in most of his other dealings, in this particular moment (whose sadomasochistic qualities seem heightened by Connie's scant clothing, Vic's kiss, and Connie's whispered "hit me,"), Vic's heart seems to get in the way. However, loving Connie as he does, Vic puts aside his personal doubts and strikes her for her own good. At once both viciously brutal and hopelessly compassionate, *The Shield* suggests, Vic's is tough love taken to an extreme.

Ultimately, Vic is both more and less sensitive than the other police officers in his department, as these scenes seem to suggest. While he shuns political correctness and refuses to be sensitive to Captain

Aceveda's cultural background, it is Vic who is most connected and most friendly with the Latino and African American gang members in his precinct. It is Vic whom these gang members most trust and fear, further evidence of his conflicted character. Likewise, while Vic seems more accepting of Julien's homosexuality than many of the other officers in the department, it is Vic who uses it to blackmail and intimidate him. Both tough and compassionate, Vic Mackey seems a truly compassionate conservative—a conflicted character whose various values struggle against one another from one moment to the next.

Masculinity and the Politics of Backlash

The Sopranos began in 1999, two years prior to the events of September 11 and a year prior to the beginning of the Bush presidency. Still, the show seems to have been prophetic of the masculine ideas that would follow (in *The Shield* and elsewhere) as the American culture became more reflective about the seemingly new, sensitive masculinity it had created within the '90s. Together, Tony Soprano and Vic Mackey illustrate pieces of a backlash that really isn't one. On the one hand, these early 21st-century characters seem to have displaced the so-called new men of the '90s as these sensitive men begin to lose their cultural cachet (note, for instance, how David Schwimmer's sensitive Ross character on *Friends*—discussed in chapter 2—transforms from the show's primary focus into a more obnoxious, whining joke, evidence of this shift in the cultural imagination). By poking fun at sensitivity—as do many of Tony's outbursts in Dr. Melfi's office, or laughing in the face of multiculturalism—as do many of Vic's comments—these programs seem to pronounce the death of a caring, sensitive manhood. Similarly, magazines such as *Maxim* and television programs such as *The Man Show*, which arrived on the cultural scene at roughly the same moment as these two other programs, make explicit their own celebrations of traditional masculinity. In their objectification of women and their homage to beer, gadgets, sports, and other "manly things," these and similar sources have suggested that a new traditional masculinity has arrived. Here are images of masculinity, it seems, fed up with all of that new-age nonsense of the past.

However, as much as Tony and Vic explicitly celebrate the masculinity of the past, the demons that both programs give them to fight suggest that these characters—like the culture more generally—are still uncomfortably negotiating the meanings of manhood and the perceived tension between toughness and sensitivity. In the process, these characters repeat many of the anxieties at play within the seemingly sensitive men that precede them. For instance, Tony and Vic have a host of affairs with stereotypically attractive women. Here, like the little men of the '90s sitcom discussed in chapter 2, they reclaim a traditional masculine stereotype, offsetting their sensitivity with a hypersexuality that demonstrates their continuing willingness to objectify and exploit women. The conflicts between Vic and Julien make homophobia an explicit story line within *The Shield*, repeating a tension that plays throughout the men of the '90s, discussed in the previous chapters. While both of these characters might seem decidedly more complex and interesting than many of those discussed in the previous chapters, the continued repetition of these same figures and stereotypes demonstrates the limited resources currently available for representing masculinity. These television programs, like much of the mainstream culture more generally, seem to fall back on safe, repeated notions of manhood, continuing many of the problems inherent in this traditional masculinity.

These programs illustrate a more general conception of masculinity that seems to have taken hold at the turn of the century. In her best-selling book *Stiffed: The Betrayal of the American Man* (1999), Susan Faludi (who won earlier acclaim with her book *Backlash: The Undeclared War against American Women*) captures a sense of confusion among men of the late '90s. The various men she interviews and discusses—like Tony Soprano himself—feel that the promise of masculinity has somehow passed them by. Similarly, as stereotypically hypermasculine as shows like *The Man Show* or its less popular twin *The X Show* seem to be, they also depict a profoundly ironic, broken vision of masculinity. Hosted by comedians such as Adam Corolla and Jimmy Kimmel—who, like Tony Soprano and Vic Mackey, are overweight, seemingly regular guys rather than stereotypical hypermasculine studs—the shows' masturbation jokes and slow-motion bouncing breast shots seem sad attempts to recapture some dream of traditional masculinity. Here, it seems, are forty-year-old frat boys attempting to hold on to their youth.

At a time of such confusion, it may not be surprising that criminals and dirty cops have become the century's heroic men. If the culture seems to some to have become too sensitive or too civilized, then who better to stand up for the American man than the hero for whom the rules of society do not apply? In such a climate, Arnold Schwarzenegger's accusations of sexual misconduct may have actually helped him in his bid to be the governor of California. Framing him as a nonpolitically correct rogue, Schwarzenegger's accusers may have unwittingly highlighted his hypermasculine heroics. In a comparable manner, George W. Bush's willingness to buck the United Nations and the rest of the world in pursuing wars in Afghanistan and Iraq seems to make him a strong example of this particular vision of manhood. Like Tony Soprano and Vic Mackey, George W. Bush seems a unilateralist who flaunts his own toughness at every opportunity. While this conflicted vision of masculinity rehashes much of the tensions and anxieties of the '90s, the increasingly schizophrenic nature of Tony and Vic offers a still more profoundly troubling vision, celebrating an extreme violence apparently ameliorated by even a modicum of compassion.

Facing up to Masculinity

Ideas about masculinity inflect discussions about September 11 and play an important role within the "Oughts" more generally. Here, the conflicted visions of manhood that permeate much of the '90s become all the more pronounced. While there is nothing intrinsically wrong with having masculine heroes, the ways in which we envision this heroism have consequences for our views of the world and of the various people inhabiting it. The masculine heroics of the '90s feigned a new, more sensitive masculinity all the while repeating a host of anxieties associated with traditional manhood. The masculinity imagined at the turn of the 21st century feigns a departure from this manhood of the '90s, only to recreate its anxieties and conflicts in a still more hyperbolic form.

During the '90s, the idea of the crisis of masculinity served a number of different purposes—from allowing the Clinton administration to depict him as a sensitive but tough new man to opening the door for Kenneth Starr's rebuke of the hypersexual, hyperemotional Clinton, from selling Steven Seagal movies to selling memberships in the Promise Keepers. The dawn of the 21st century has demonstrated still more manifestations of this sense of masculine crisis. Here, it opens up arguments for increased sensitivity in our communications at the same time that it sanctions brutal violence and bloodshed. It captures the profound pain and strength of a group of firefighters at the same time that it celebrates the cunning and toughness of criminals and dirty cops. Exploring these moments through the lens of gender theory allows us to begin to unravel their invisible hold on our contemporary culture, disrupting the propagandistic cultural work these ideas continue to perform.

Notes

1. The Boy Scouts of America, for instance, developed around the turn of the century when anxious fathers worried that their sons were spending too much time with their female teachers and mothers.

2. According to a review in the *Houston Chronicle*, "no one messes with this Andy Sipowicz on steroids. You might not like his methods, but he gets results."

Conclusion

George W. Bush's May 2003 landing aboard a navy aircraft carrier offered a conflicted image of the president's masculinity. While Bush looked the part of a fighter pilot in his green flight suit, the widely circulated images of the landing opened him up to a variety of attacks as well. Calling attention to Bush's own privileged background, a number of people saw this scene as an ironic reflection on Bush's relatively insulated military past. One anti-Bush website created its own version of the landing, editing the images into a video entitled "Triumph of the Wimp," a hardly subtle take-off on the Leni Riefenstahl film of the 1930s. Similarly, in his own bid for the presidency, John Kerry would comment:

> Our opponents say they want to campaign on national security. Well, I know something about aircraft carriers for real. And if George Bush wants to make national security the central issue of this campaign, we have three words for him we know he understands: Bring it on!

The same image that the Bush campaign chose as a display of Bush's heroic machismo (the banner "Mission Accomplished" pronounced the success of Bush's military strategies) was to others a sign of his failure. While one interpretation saw Bush as a successful commander-in-chief, another saw him as an impotent failure pretending to be a soldier.

As with a variety of strategies used by the Clinton administration before him, Bush's attempts here illustrate some of the conflicted purposes to which these masculine ideals can be put. While a presidential candidate can be framed as too excited, as was Howard Dean following his speech at the 2004 Iowa caucuses, so he can also be framed as not excited enough, as a variety of comments about 2000 presidential candidate Al Gore's "stiffness" illustrates. A presidential candidate can come off as too military focused, as some of George W. Bush's opponents suggest in framing him as an out-of-control cowboy. But he can also come off as not military minded enough—another criticism made against Bush in various references to his relatively sparse military background. While a president can be too sensitive, he can also be not sensitive enough, both of which are criticisms made of Clinton throughout his presidency.

As such a conflicted cultural ideal, masculinity is both difficult to understand and easy to wield. Although the masculine ideals of the '90s and early 21st century demonstrate a variety of paradoxes and contradictions, this does not prevent media producers from employing them to sell movie tickets or political pundits from using them to promote or discredit presidential candidates. On the contrary, their conflicted status makes these ideas about manhood all the more rhetorically useful. If ideas like cowardice, heroism, bravery, toughness, and sensitivity are ill defined, then ideologues are all the more free to throw these terms around without having to carefully defend them. In a culture in which calls for masculinity are both highly prevalent and highly ambiguous, these propagandists have a ready-made screen behind which to promote a variety of ideas and sell a variety of products.

For these reasons, the ways in which a culture understands and discusses masculinity will have important consequences for a variety of cultural, social, and political issues. As conflicted as the mainstream vision of '90s masculinity seems to be, it still manages to reinforce a variety of cultural anxieties, magnifying problematic ideas about race, class, and sexuality in particular. As sensitive as this vision of masculinity may appear relative to some past notions of manhood, it still manages to champion a conception of the unblushing male in America. Presenting white, middle-class, heterosexual masculinity—in the character of such new men as Fox Mulder, Jack Dawson, Ross Geller, and Bill Clinton—as the ultimate measure of "new maleness," the men of the '90s resurrect the vision of manhood they presume to leave behind.

These various masculine ideals are both specific to particular historical periods and cyclical. Leonard DiCaprio's Jack Dawson might have seemed an unlikely hero for the 1950s United States, though James Dean's character "Jim" in *Rebel without a Cause* illustrates some similar conflicts. The late '90s film *Fight Club* (based on the novel of the same name) sees Edward Norton's character Jack, who spends his time buying and dreaming about IKEA furniture, fantasize his own anticonsumer alter ego, Tyler Durden (Brad Pitt), then watches as these two characters do battle. Here, before the birth of the metrosexual male consumer, *Fight Club* offered a specific statement about the feminizing effects of mass consumption and of the struggle of American men to come to terms with this. While this might seem unique to this particular culture and period of time, Jean Luc Godard's 1965 film *Pierrot le Fou* paints a similar picture of French masculinity in the mid 1960s. Here, the film's main character is split between two personae, the good bourgeois father, Ferdinand, and the wild gangster, Pierrot. The end of the film, like *Fight Club*, sees one of these characters murder his alter ego.[1]

Similarly, while the '90s men discussed in previous chapters illustrate a variety of specifically '90s characteristics, they also share things in common with figures at different moments in time. For instance, Hawkeye Pierce of the 1970s program *M*A*S*H*, seems to have much in common with the so-called new men of the '90s. A sensitive man who despises war but still chases a variety of women, Hawkeye's masculinity is conflicted in a variety of ways. Likewise, as another moment in which feminism played an important public role, the more general gender climate of the 1970s shares certain characteristics with that of the 1990s, creating a similar set of discussions and a similar call for male sensitivity. While these similarities are important, so are the specific ways in which these various images depict their particular cultural moment; Hawkeye's character offers a specific critique of the Vietnam War just as the men of the '90s respond to particularities of this cultural moment. Ultimately, we need to understand these various figures as both unique manifestations of their particular historical moments and pieces of a larger history of American manhood.

In exploring these ideas, we can begin to understand some of the bizarre ways in which masculinity works in our everyday culture. Doing so seeks to make visible an identity that too often goes unexplored. Exploring the ways in which dominant conceptions of masculinity are negotiated and renegotiated from one moment to the next offers a means of understanding some of the ways in which American manhood connects with other dominant identities: whiteness, heterosexuality, middle-class status. Equally important, however, are explorations of the ways in which our culture depicts the alternative masculinities of sexual, ethnic, and other minority groups, as well as how female masculinity (Halberstam, 1998) is itself represented. For this reason, the analyses offered here depict but one small piece of the masculinities of the cultural moments I explore. Still, because these dominant depictions will have consequences for the ways in which these other masculinities are depicted, I hope my discussions can help in these other kinds of studies as well.

Ultimately, my analyses here aim to trouble our cultural understandings of manhood. In this regard, I have made strategic use of the concept of the crisis of masculinity, as have the various politicians, media producers, gender scholars, and writers I have explored throughout. In each of the previous chapters, I have worked to highlight the contradictory nature of a variety of American male images. In doing so, I have hoped to discover some of the more bizarre features of American masculinity and to demonstrate the ways in which these oddities informed a variety of a public discussions during the 1990s. I've also hoped to paint these ideas about masculinity with an indelible mark of their utter weirdness, parodying a set of concepts that seem so natural and taken-for-granted. My goal here is not only to describe a set of empirically present contradictions at work within the media and elsewhere but to make more difficult the invisible manipulation that these ideas about masculinity allow. If we continually view masculinity as a conflicted, cultural oddity, then we may be less likely to be easily moved when politicians and others use these ideas to promote particular actions and ways of thinking.

Owing to the dynamic nature of our cultural conceptions of masculinity, this must be an ongoing conversation. As I've suggested throughout, each new idea about masculinity opens up new possibilities for thinking about manhood. These possibilities become general cultural resources available for a variety of different aims, as has been the case for the concept of the crisis of masculinity during the 1990s and beyond. A number of scholars have argued that emphasizing that culturally constructed nature of masculinity can unhinge some of its cultural power. The Promise Keepers, on the other hand, find ways in which the language of social construction can benefit their conservative reclamation of traditional masculine and family values. In the book, *Real Men Worship* (Boschman, 1996), the thoroughly right-wing Promise Keepers argue that men's "cultural conditioning" has had a negative impact on their worship experiences, using the cultural construction of masculinity as a trope in favor of Promise Keeping. These ideas will continue to mutate and transform as long of the culture finds value in masculinity as a concept. They need to be constantly struggled with and thought about by scholars and the general public alike.

In addition to the kinds of images explored here, media and gender scholars also need to think about the political economy of American masculinity, investigating the ways in which ideas about manhood reflect particular economic conditions and serve specific economic and political aims. During the 1980s, the hard-bodied images of Sylvester Stallone and others reflected Reagan-era politics, but they also demonstrated a larger economic change within Hollywood. At that time, Hollywood producers made a concerted effort to sell their films around the world and these hypermasculine action heroes seemed perfect for that purpose. Lacking the nuances and depth of more complicated figures, these '80s heroes—and their requisite explosions and gunfights—were ready made for a global market that could easily translate them into other cultures. Similarly, the hypersensitive care with which a variety of media covered a variety of issues immediately after September 11 was likely connected to lobbying efforts undertaken by several large media conglomerates during this time. With the radio giant Clear Channel soliciting the government to decrease media ownership restrictions, the company's September 11 sensitive play lists (which excluded antiwar songs, as well as a host of other apparently insensitive music) likely sought to win them the favor of the current administration.

Additionally, gender scholars, and media and cultural studies scholars more generally, need to continue developing tools for exploring the ways in which particular anxieties and emotions become inscribed and reiterated in different cultural discourses—from films and television programs to presidential speeches and radio play lists. The analyses in the previous sections suggest that many conceptions of masculinity may be too complex to fit under the fairly standard cultural studies term "ideology." Rather than a coherent ideological picture of American manhood, the '90s and early 21st-century masculinities explored in the previous chapters present conflicted and often contradictory ideas about maleness that variously serve to uphold and challenge traditional masculine ideals. Likewise, these '90s negotiations are more complex than simple matters of the mainstream culture co-opting or "incorporating" alternative ideas about masculinity (Gramsci, 1971; Williams, 1991). Rather than simply pacifying these alternative identities, the '90s films and television programs discussed here take full advantage of their multiple possibilities, using them, as does the late '90s film *Man on the Moon*, to recreate traditional masculine values. In order to address these interesting cultural moments and their attendant cultural anxieties, scholars must complicate their discussions to explore the complexities of emotionology (Stearns and Stearns, 1985; Malin, 2001) in addition to ideas about ideology more generally.

Exploring these complex negotiations of identity evinces the complicated pull of masculinity within our contemporary culture. Here is a contradictory cultural ideal infecting a variety of messages. President Clinton and his fellow '90s men, along with the masculine characters of the early 21st century, provide important opportunities for understanding and observing the ways in which various crises of masculinity are struggled through, as well as for intervening in this process of negotiation. Holding up these conflicted images and their concomitant anxieties seeks to both diagnose and disrupt the reproductive power of traditional masculinity, continuously working to reframe the *new possibilities* of masculinity in ways antithetical to these conventionalized histories.

Notes

1. However, *Pierrot le Fou* suggests a less happy ending than does *Fight Club* through its reference to Edgar Allan Poe's story "William Wilson," which Ferdinand recounts earlier in the film. "You have conquered, and I yield" Poe's story ends. "Yet, henceforward art thou also dead—dead to the World, to Heaven and to Hope! In me didst thou exist—and, in my death, see by his image, which is thine own, how utterly thou hast murdered thyself."

References

Alexander, S., and L. Karaszewski. (1999). *Man on the moon: The shooting script*. New York: New Market Press.

Alter, J. (September 21, 1998). Between the lines: Spinning out of sinning. *Newsweek*, 45.

Althusser, L. (1971). *Lenin and philosophy and other essays* (B. Brewster, Trans.). London: New Left Books.

Andersen, K. (June 21, 1993). Are Beavis and Butthead arty? *Time*, 75.

Ansen, D. (February 23, 1998). Our Titanic love affair. *Newsweek* 131, 58–62.

Bakhtin, M. (1984). *Rabelais and his world* (H. Iswolsky, Trans.). Bloomington: University of Indiana Press.

Bellon, J. (1999). The strange discourse of *The X-Files*: What it is, what it does, and what is at stake. *Critical Studies in Mass Communication 16*(2), 136–154.

Bennett, J. (March 23, 1998). Clinton packs up his care and woe to trot the globe. *New York Times*, section A, 1.

Berger, M., B. Wallis, and S. Watson (Eds.). (1995). *Constructing masculinity*. New York, London: Routledge.

Berlant, L. (1997). *The queen of America goes to Washington City: Essays on sex and citizenship*. Durham, NC: Duke University Press.

Bernardi, D. (1998). *Star Trek and history: Race-ing toward a white future*. New Brunswick, NJ: Rutgers University Press.

Biskind, P., and B. Ehrenreich. (1987). Machismo and Hollywood's working class. In D. Lazere (Ed.), *American media and mass culture: Left perspectives* (pp. 201–215). Berkeley and Los Angeles: University of California Press.

Bly, R. (1990). *Iron John: A book about men*. Reading, MA: Addison-Wesley.

Blythe, H., and C. Sweet. (1986). Coitus interruptis: Sexual symbolism in "The Secret Life of Walter Mitty." *Studies in Short Fiction* 23(1), 110–113.

Boschman, L. (1996). *Real men worship.* Ann Arbor, MI: Servant Publications.

Bourdieu, P. (1984). *Distinction: A social critique of the judgement of taste* (R. Nice, Trans.). Cambridge, MA: Harvard University Press.

Bourdieu, P., and J. Passeron. (1994). *Reproduction in education, society and culture.* London, Newbury Park, New Delhi: Sage Publications.

Brenton, M. (1966). *The American male.* New York: Coward-McCann.

Burros, M. (December 23, 1992). Bill Clinton and food: Jack Sprat he's not. *New York Times,* Section C, 1.

Butler, J. (1990). *Gender trouble: Feminism and the subversion of identity.* New York and London: Routledge.

———. (1993). *Bodies that matter: On the discursive limits of sex.* New York and London: Routledge.

———. (1997). *Excitable speech: A politics of the performative.* New York, London: Routledge.

Calhoun, C. (Ed.). (1994). *Habermas and the public sphere.* Cambridge, MA, and London: The MIT Press.

Campbell, R. (1991). *60 minutes and the news: A mythology for middle America.* Urbana: University of Illinois Press.

Carey, J. (1977). Mass communication research and cultural studies: An American view. In J. Curran, M. Gurevitch, and J. Woollacott (Eds.), *Mass communication and society* (pp. 409–425). London: Sage.

———. (1992). *Communication as culture.* New York and London: Routledge.

Cheatham, G. (1990). The secret sin of Walter Mitty. *Studies in Short Fiction* 27(4), 608–610.

Christon, L. (May 18, 1985). The staging of Andy Kaufman's "greatest" stunt. *Los Angeles Times,* 7.

Church, G. J. (January 27, 1992). Is Bill Clinton for real? *Time,* 14–24.

Cloud, D. (1998). *Control and consolation in American culture and politics: Rhetorics of therapy.* London: Sage.

Coe, S. (1990). *Dream On* could be HBO sleeper. *Broadcasting* 119(11), 75–76.

Connell, R. W. (1993). The big picture: Masculinities in recent world history. *Theory and Society* 22, 597–623.

Cuomo, C., and K. Hall. (1999). *Whiteness: Feminist philosophical reflections.* Lanham, MD: Rowman & Littlefield.

Curry, T. (1991). Fraternal bonding in the locker room: A profeminist analysis of talk about competition and women. *Sociology of Sport Journal 8*(2), 119–135.

Darsey, J. (1981a). "'Gayspeak': A response." *Gayspeak: Gay male and lesbian communication* (pp. 58–67). New York: Pilgrim Press.

———. (1981b). "From 'Commies' and 'Queers' to 'Gay is good.'" *Gayspeak: Gay male and lesbian communication* (pp. 224–247). New York: Pilgrim Press.

———. (1994). "Die Non: Gay liberation and the rhetoric of pure tolerance." *Queer words, Queer images: Communication and the construction of Homosexuality* (pp. 45–76). New York and London: New York University Press.

Deem, M. (1999). Scandal, heteronormative culture and the disciplining of feminism. *Critical Studies in Mass Communication 16*, 86-94.

———. (2001). The scandalous fall of feminism and "the first black president." In T. Miller (Ed.), *The Blackwell companion to cultural studies.* Oxford, UK: Blackwell.

Delgado, R., and J. Stefancic. (1997). *Critical white studies: Looking behind the mirror.* Philadelphia, PA: Temple University Press.

Deming, C. (1991). *Hill Street Blues* as narrative. In R. K. Avery and D. E. Eason (Eds.), *Critical perspectives on media and society* (pp. 240–264). New York and London: Guilford Press.

Doyle, J. (1983). *The male experience.* Dubuque, IA: W. C. Brown.

Doyle, R. F. (1976). *The rape of the male.* St. Paul, MN: Poor Richard's Press.

Dyer, R. (1997). *White.* London and New York: Routledge.

Ebert, T. L. (1996). *Ludic feminism and after: Postmodernism, desire, and labor in later capitalism.* Ann Arbor: University of Michigan Press.

Ehrenreich, B. (1983). *The hearts of men.* Garden City, NY: Anchor Press/Doubleday.

Faludi, S. (1981). *Backlash: The undeclared war against American women.* New York: Crown.

———. (1999). *Stiffed: The betrayal of the American man.* New York: Harper Collins.

Fasteau, M. (1975). *The male machine.* New York: McGraw-Hill.

Farrell, W. (1974). *The liberated man.* New York: Random House.

————. (1986). *Why men are the way they are.* New York: McGraw-Hill.

————. (1993). *The myth of male power: Why men are the disposable sex.* New York: Simon & Schuster.

Fiske, J. (1987). *Television culture.* London and New York: Routledge.

————. (1989). *Understanding popular culture.* London and New York: Routledge.

————. (1991). Television: Polysemy and popularity. In R. K. Avery and D. E. Eason (Eds.), *Critical perspectives on media and society* (pp. 346–364). New York and London: Guilford Press.

Foucault, M. (1995). *Discipline and punish* (A. Sheridan, Trans.). New York: Vintage Books.

Frankenberg, R. (Ed.). (1997). *Displacing whiteness: Essays in social and cultural criticism.* Durham, NC: Duke University Press.

Friedan, B. (1963). *The feminine mystique.* New York: W. W. Norton.

Garber, M. (1992). *Vested interests: Cross-dressing and cultural anxiety.* New York and London: Routledge.

Gardner, J. (May 2, 1994). Leave it to Beavis. *National Review,* 60–63.

Gitlin, T. (1983). *Inside prime time.* New York: Pantheon Books.

Goffman, E. (1963). *Stigma: Notes on the management of spoiled identity.* Englewood Cliffs, NJ: Prentice-Hall.

————. (1979). *Gender advertisements.* Cambridge, MA: Harvard University Press.

Goldberg, S. (1974). *The inevitability of patriarchy.* New York: Morrow.

Goldman, A. (1981). *Elvis.* New York, St. Louis, San Francisco, Hamburg, Mexico, Toronto: McGraw-Hill.

Gopnick, A. (1994). The great deflater. *New Yorker 70* (June 27–July 4), 168–173.

Gramsci, A. (1971). *Selections from the prison notebooks of Antonio Gramsci.* New York: International Publishers.

Gronbeck, B. (1997). Character, celebrity, and sexual innuendo in the mass-mediated presidency. In J. Lull and S. Hinerman (Eds.), *Media scandals: Morality and desire in the popular culture marketplace* (pp. 122–142). New York: Columbia University Press.

Halberstam, J. (1998). *Female masculinity.* Durham, NC: Duke University Press.

Hall, S., and T. Jefferson (Eds.). (1975). *Resistance through rituals: Youth subcultures in post-war Britain.* London: Routledge.

Hall, S. et al. (1978). *Policing the crisis: Mugging, the state and law and order.* New York: Holmes and Meier.

Harrison, T., Projanksy, S., Ono, K., Helford, E. (Eds.). (1996). *Enterprise zones: Critical perspectives on Star Trek.* Boulder, CO: Westview Press.

Hebdige, D. (1979). *Subculture: The meaning of style.* London and New York: Routledge.

Hefner, H. (1953). Introduction. *Playboy,* December 1953, 3.

Hill, M. (1997). *Whiteness: A critical reader.* New York: NYU Press.

Hirshey, G. (1996). The comic angst of David Schwimmer. *GQ 66*(3), 232–237.

Hoggart, R. (1992). *The uses of literacy* (Originally published in 1957). New Jersey: Transaction Publishers.

Hopkins, J. (1971). *Elvis.* New York: Simon & Schuster.

Janson, S. C. (1994). The sport war metaphor: Hegemonic masculinity, the Persian Gulf War, and the New World order. *Sociology of Sport Journal 11*(1), 1–17.

Jeffords, S. (1994). *Hardbodies: Hollywood masculinity in the Reagan Era.* New Brunswick, NJ: Rutgers University Press.

Jenkins, H. (1992). *Textual poachers: Television fans and participatory culture.* London and New York: Routledge.

Kahn, S. (1996). Who, me? A star? *McCall's 66*(3), 48–52.

Kaufman, A. (1994). "Things close in": Dissolution and misanthropy in "The Secret Life of Walter Mitty." *Studies in American Fiction 22*(1), 93–104.

Kellner, D. (1995). *Media culture.* New York: Routledge.

Kimmel, M. (1986). Toward men's studies. *American Behavioral Scientist 29*, 517–530.

———. (1993). Invisible masculinity. *Society 30*(6), 28–35.

———. (1996). *Manhood in America: A cultural history.* New York: Free Press.

Kolbert, E. (September 28, 1992). The governor: Clinton in Arkansas. *New York Times,* Section A, 1.

Kurtz, H. (July 9, 1992). The airways: Hitting Clinton with Flowers by wire; Operative behind Horton ad to promote call-in for "intimate" tapes. *Washington Post,* Section A, 10.

Lane, A. (1995). Style wars. *New Yorker.* 94–96.

———. (December 15, 1997). The shipping news: *Titanic* raises the stakes of the spectacular. *New Yorker 73*, 156–158.

Lipton, M. A., and T. Cunneff. (1995). Revenge of the nerd. *People Weekly 43*(6), 179–180.

Lott, E. (1997). All the king's men: Elvis impersonators and white working-class masculinity. In H. Stecopoulos and M. Uebel (Eds.), *Race and the subject of masculinities* (pp. 192–227). Durham and London: Duke University Press.

Lusane, C. (1999). Assessing the Disconnect between black and white television audiences: The race, class, and gender politics of *Married …with Children. Journal of Popular Film and Television, 27*(1), 12–20.

Lutz, T. (1999). *Crying: The natural and cultural history of tears.* New York: W. W. Norton.

Malin, B. (2001). Communication with feeling: Emotion, publicness, and embodiment. *Quarterly Journal of Speech 87*(2), 216–235.

Maraniss, D. (October 4, 1992). Clinton's approach to racial matters: Conflicting impulses. *Washington Post*, Section A, 1.

Marcus, G. (1991). *Dead Elvis.* New York, London, Toronto, Sydney, Auckland: Doubleday.

Marin, P. (July 8, 1991). The prejudice against men. *Nation 253*(2), 46–51.

Marin, R. (1995). Triumph of a coffee bar Hamlet. *Newsweek 125*(17), 68–69.

———. (April 29, 1996). Nuking the nuclear family. *Newsweek,* 70.

Mathias, B. (December 14, 1992). Style plus: Focus, the sensitive touch, and the advent of a hugging president. *Washington Post*, Section B, 5.

McCartney, B. (1992). *What makes a man? 12 promises that will change your life.* Colorado Springs, CO: Navpress.

McConnell, F. (January 14, 1994). Art is dangerous: *Beavis and Butthead,* for example. *Commonweal,* 28–30.

Means, S. (1995). Born to be mild. [On line]. http://www.film.com /film-review/9565/27/default- review.html.

Mellencamp, P. (1990). Critical and textual hypermasculinity. *Logics of television: Essays in cultural criticism* (pp. 156–172). London: BFI Publishing.

Mercer, K. (1994). *Welcome to the jungle: New positions in Black Cultural Studies.* New York and London: Routledge.

Messner, M. (1993). "Changing men" and feminist politics in the United States. *Theory and society 22*(5), 723–737.

Meštrovic, S. (1997). *Postemotional society.* London: Sage Publications.

Miles, I. (1989). Masculinity and its discontents. *Futures 21*(1), 47–59.

Miller, M. C. (1987). Deride and conquer. In T. Gitlin (Ed.), *Watching television* (pp. 183–246). New York: Pantheon Books.

Morrow, L. (September 14, 2001). The case for rage and retribution. *Time*, 48.

———. (November 19, 2001). Has your paradigm shifted? *Time*, 152.

Nakayahama, T., and Martin, J. (1998). *Whiteness: The communication of social identity*. London: Sage Publications.

New York Times. (May 1, 1998). Mr. Clinton's awkward answers. 26.

Pfeil, F. (1995). *White guys: Studies in postmodern domination and difference*. London and New York: Verso.

Pollitt, K. (March 30, 1998). Women and children first. *Nation*, 9.

Powers, J. (January 1998). People are talking about movies. *Vogue*, 78–80.

Rebeck, V. (June 27–July 4, 1990). Recognizing ourselves in the Simpsons. *Christian Century*, 618–622.

Reid, T. R. (May 15, 1998). British media stress sex rather than summit. *Washington Post*, 30.

Rich, F. (April 4, 1998). The Viagra kid. *New York Times*, Section A, 13.

Robbins, B. (Ed.). (1993). *The phantom public sphere*. London and Minneapolis: University of Minnesota Press.

Rodman, G. (1996). Elvis after Elvis: The posthumous career of a living legend. New York: Routledge.

Rogin, M. (1996). Blackface, white noise: Jewish immigrants in the Hollywood melting pot. Berkeley: University of California Press.

Ross. A. (1995). The great white dude. In M. Berger, B. Willis, and S. Watson (Eds.), *Constructing masculinity* (pp. 167–175). New York: Routledge.

Rotundo, E. A. (1993). *American manhood*. New York: Basic Books.

Roy, B. (1994). *Some trouble with cows: Making sense of social conflict*. Berkeley, Los Angeles, London: University of California Press.

Rubin, P. (April 27, 1998). Family man. *New Republic*, 14.

Russell, H. (1994). Crossing games: Reading black transvestism at the movies. *Critical Matrix: The Princeton Journal of Women, Gender, and Culture 8*(1), 109–125.

Segal, L. (1993). Changing men: Masculinities in context. *Theory and Society 22*(5), 625–641.

Shweder. R. A. (January 9, 1993). What do men want? A reading list for the male identity crisis. *New York Times Book Review*, 3.

Simms, P. (June 29, 1989). Smellovision. *Rolling Stone*, 30.

Sloop, J. (2000). Disciplining the transgendered: Brandon Teena, public representation, and normativity. *Western Journal of Communication, 64*(2), 165–189.

Stearns, P., and C. Stearns. (1985). Emotionology: Clarifying the history of emotional standards. *American Historical Review 90*(4), 813–836.

Stecopoulos, H., and M. Uebel (Eds.). (1997). *Race and the subject of masculinities*. Durham and London: Duke University Press.

Stephanopoulos, G. (September 21, 1998). In the bunker: Why he'll try to fight. *Newsweek*, 37.

Stiglitz, J. (April 17 and 24, 2000). What I learned at the world economic crisis. *New Republic* 4(448 and 449), 56–60.

Stillman, D. (February 27, 1994). The trouble with male bashing. *Los Angeles Times Magazine*, 32–38.

"Talk of the town." (1998). *New Yorker, 74*(3), 31–39.

Tillotson, K. (June 13, 1999). Steven Seagal: Personal impression leaves no bruises. *Minneapolis Star-Tribune*, F9.

Toner, R. (September 14, 1992). The 1992 campaign: Political memo: Anxious in his lead, Clinton fights to run race his way. *New York Times*, Section A, 1.

Tucker, R. C. (Ed.). (1978). *The Marx-Engels reader*. New York, London: W. W. Norton.

Uebel, M. (1997). Men in color: Introducing race and the subject of masculinities. In H. Stecopoulos and M. Uebel (Eds.), *Race and the subject of masculinities* (pp. 1–14). Durham and London: Duke University Press.

Vernon, E. (Ed.). (1976). *Humor in America*. New York: Harcourt Brace Jovanovich.

Von Drehle, D. (March 7, 1992). Letter from the campaign trail: Clinton's political persona blends redneck, policy nerd. *Washington Post*, Section A, 1.

Vonnegut, K. (1961). *Mother night*. New York: Dell Publishing.

Walczak, L., et al. (March 23, 1992). Is it still a horse race? *Business Week*, 26.

Wallace, B. (July 20, 1992). Down home in Clinton's Dixie. *Maclean's*, 27.

Warner, M. (1994). The mass public and the mass subject. In C. Calhoun (Ed.), *Habermas and the public sphere* (pp. 377–401). Cambridge, MA, and London: MIT Press.

Wiegman, R. (1995). *American anatomies: Theorizing race and gender.* Durham and London: Duke University Press.

Wilcox, R. (1996). Miscegenation in *Star Trek: The Next Generation.* In T. Harrison, S. Projanksy, K. Ono, and E. Helford, (Eds.), *Enterprise zones: Critical perspectives on Star Trek* (pp. 69–92). Boulder, CO: Westview Press.

Williams, R. (1958). *Culture and society: 1780–1950.* New York: Columbia University Press.

———. (1977). *Marxism and literature.* Oxford and New York: Oxford University Press.

———. (1991). Base and superstructure in Marxist cultural criticism. In C. Mukerji and M. Schudson (Eds.), *Rethinking popular culture* (pp. 398–404). Berkeley, Los Angeles, Oxford: University of California Press.

Williams, S., and G. Bendelow. (1998a). *The lived body: Sociological themes, embodied issues.* London and New York: Routledge.

———. (Eds.). (1998b). *Emotions in social life: Critical themes and contemporary issues.* London and New York: Routledge.

Willis, P. (1977). *Learning to labor: How working-class kids get working-class jobs.* New York: Columbia University Press.

Zizek, S. (1989). *The sublime object of ideology.* London: Verso.

Index